Had I Known

Had I Known

A Memoir of Survival

Joan Lunden

with Laura Morton

HARPER LUXE

An Imprint of HarperCollinsPublishers

HarperCollins books may be purchased for educational, business, or sales promotional use. For information, please e-mail the Special Markets Department at SPsales@harpercollins.com.

FIRST HARPERLUXE EDITION

HarperLuxe™ is a trademark of HarperCollins Publishers

Library of Congress Cataloging-in-Publication Data is available upon request.

ISBN: 978-0-06-241687-2

15 ID/RRD 10 9 8 7 6 5 4 3 2 1

I dedicate this book to my amazing family, my friends, and all those who reached out to me day after day on social media, sending me their prayers and well wishes. You have all provided me with healing strength.

While I'm not always forthcoming with expressing how I feel with my emotions, it's necessary to say a thank-you. My love runs deep, and my heart is full from how you rallied around me in my time of great need. I will admit to you only now that yes, I needed you. You have always been there for me when I've needed you, even if I never said it. And yet never more unconditionally than how you gathered around me during these past nine months. Your strength, support, and unwavering love did not go unnoticed and certainly didn't go unappreciated. Your strength and

encouragement are what helped me remain so strong—
I could not have made it through the fire without all of
you.

This is the story and my personal journey of how
I faced my breast cancer diagnosis and subsequent
treatment head-on. Yup, it's the good, the bad, and
the bald, ugly truth.

Breast cancer is not a one size fits all disease. It
requires great personalization for treatment and therefore no two battles are the same. If you think you
might have breast cancer or are currently in treatment,
it's critically important that you seek your own medical
advice for proper diagnosis and treatment. The information I offer throughout the pages of this book merely
details the path I chose to follow for my personal treatment. There are many roads to walk, and this is the
one I took. I am not a medical professional and therefore the advice and information I share throughout this
book are in no way intended to be offered as medical
advice.

For whatever reason, it has never been easy for me
to accept help from anyone, family or friends, so
when I was diagnosed with breast cancer, I wasn't
sure how or even if I would allow myself to accept the

unconditional love and kindness you all showed. It didn't take a dance with death to know that I had an incredible partner in my husband, Jeff. I have had Jeff by my side throughout this journey, and in my head: Each time I felt down or stressed out, he would remind me of my personal fortitude and the enormous amount of love and support I had. Jeff has truly been my anchor through it all—he's shown such amazing love, loyalty, and patience. I especially appreciate that last one, his patience.

Of course, none of this has surprised me. Jeff is one of the most loyal, caring, and conscientious people I've ever known. When you meet Jeff, it's hard not to be bowled over by his presence. He's incredibly charismatic, and what can I say, he's a tall, handsome, athletic guy. And you would think that would be enough, wouldn't you? Yet those characteristics that one would look for in a mate take a backseat to his character, his loyalty, his compassion, his convictions, and his tenderness. I remember the first day I ever saw Jeff, at a deli in Rye Brook, New York. Ironically, I was having lunch with my friend Laura (and the coauthor on this book) and my daughter Sarah, who was ten at the time. I looked up and saw him standing in the doorway, looking around the restaurant, and suddenly, our eyes met. He flashed me that thousand-watt smile of his.

I turned to Sarah and said, "How can I meet a nice guy like that?"

To which Sarah said, "How do you know he's nice?" And being ten, she added, "Go say hi and meet him."

I told her it didn't work that way.

"Why not?" she asked.

I explained that girls don't just walk up to boys and say, "I'd like to get to know you."

She shot back, "Why not? That's stupid!"

Okay, so maybe she had me there.

Jeff kept looking at me throughout lunch, and I'll admit, I kept sneaking looks at him, too. The funniest part was we kept catching each other looking. I'd smile but turn away, embarrassed, like a schoolgirl who'd gotten caught staring at her crush.

"What do you like about him, Mom?" Sarah asked.

I told her I thought he looked confident but not arrogant; that he looked very kind; and that he had a smile that could light up the Empire State Building.

It was right about then that Jeff boldly walked to our table and introduced himself and ultimately asked for my number so he could take me out to dinner sometime. I was single, but clumsy single. I was also not so quick to give out my phone number, even though I desperately wanted to. Laura, being a great friend, figured out right away that I was too shy and scared to give him my number, so she grabbed a pen from her bag and wrote my office number on a piece of paper and handed it to him.

That was a Saturday.

On Monday morning, he had already called the office by the time I was off the air from *Good Morning America*. From our first date, I believe we both knew that not only did we feel immense chemistry, but we greatly admired each other and what we stood for in life. So I've always known he was special, though only this journey has shown me just how special.

And the apple doesn't fall far from the tree.

My mother-in-law, Janey, often jokingly reminds me, "I've trained him well!"

Yes, you did, Janey!

Of course, her husband, Donny, helped, too.

Going through chemo produces something known as "chemo brain"—it's a kind of fog that gets in the way of concentrating and hearing things the first time people tell you something. One day Jeff laughingly equated living with my chemo brain with a woman putting up with a man's selective hearing. I love that analogy. Don't tell anyone, but I may have to milk this one for a while.

I have always had wonderful relationships with my daughters Jamie, Lindsay, and Sarah. I'm so proud of the lovely women they've become. However, a disease like cancer can connect and bond a family in such miraculous ways that relationships deepen, bringing out the very best in everyone. It can also send them running for the hills out of sheer panic or, worse, break them apart.

My older daughters have been there for me in ways I could have only hoped and prayed for but never fully imagined. They were by my side for every doctor's visit and treatment, regardless of what they had on their schedules and despite my assurance that I didn't need them to be. They checked in with me daily, just to be sure I was keeping my glass half full (both philosophically and literally!).

When Lindsay gave birth to my first grandchild, her beautiful little girl, Parker Leigh, in late August 2014, she gave me the most precious reason to maintain the "warrior mode" I had gone into from pretty much the moment I was diagnosed. Looking into the eyes of my daughter's beautiful baby girl, I knew I had to fight harder than ever to beat my disease. There were so many days ahead for all of us to celebrate the joys of life—not the tragedy of possible death.

When Lindsay gave birth, she took three months' maternity leave from her role as vice president at my production company. I was wonderfully blessed to have

Lindsay's younger sister, Sarah, fill her shoes. At the time, Sarah was living and working in television production in Los Angeles but was more than willing to come back east to help out. It was no small request! She was just hitting her stride in L.A. when I called to ask if she would consider putting her life on hold—assuring her that it would be only for a short time, until Lindsay was able to come back to work. I knew it was a bold move for Sarah to leave California and her job on such short notice. But I didn't know who else to turn to.

Jamie was immersed in her career in New York City as a successful publicist. She was also newly married, and while she and her husband, George, often come to our house in Connecticut to visit, I knew that with her demanding schedule, her ability to be with me through the ordeal would be limited, and I didn't want to make her feel guilty about that—there was nothing she could do about it, and I completely understood.

I hadn't planned to replace Lindsay when she went on maternity leave, but when you added cancer to the picture, I definitely needed help. I'd barely eked out my words before I heard my stoic and supportive daughter say, "Absolutely, Mom, I will be there for you, whatever you need me to do." Thankfully, Sarah didn't hesitate. Not even a little bit. She must have sensed my worry and need even if I didn't display them. In no time, she was home, at my side, and at the helm of my company. I don't know how I can ever thank her for what she did for me; I know it played a major part in my battle and recovery.

It is one thing when family comes to your rescue in the middle of a crisis, but then there are those special friends, the ones you know, without a doubt, will be there for you no matter how ugly things get. Marlene Dietrich once said it's the friends you can call at four A.M. who matter the most. Well, I found out who those friends are—fast. Like Elise Silvestri, who was my first personal assistant at *Good Morning America* in 1980, when I first stepped into the cohost role, and has worked with me on countless projects ever since; and Jill Seigerman, who also worked as my personal assistant at *GMA* and in the years after. These are deep friendships that are exemplified in such giving and dedicated ways, and they have continued to serve as a stronghold in my life. These women were the very first to be called—they know me so well that they understand how I'm feeling now. I knew they would be there to help me make the tough and bold decision of going public, and they would also be there to hold me up when the time came to help me pick out wigs and to tell me I still looked like me . . . which would definitely be needed.

Quite honestly, as soon as I heard the words "You have cancer" I was certain I would want to document my story and ultimately share this journey, and I knew there was only one person I would do that with: Laura Morton. She is a longtime friend and was already coauthor of several of my other books, and we had a history of writing together—a complete trust—and I knew this would be a vulnerable mission for me, a time in my life when authenticity would rule. I called her in the very early hours of this crisis and asked her, maybe begged her, to be on board with this memoir, for there is no one else with whom I would have entrusted the telling of this journey.

Then there are my longtime trusted advisers: my attorney of over thirty years, Marc Chamlin; my longtime accountant, Richard Koenigsberg; of course, my dear friend and colleague Charlie Gibson; and the handful of other lifelong friends with whom I always eagerly look forward to sharing joys and who are the few with whom I am willing to share my sorrows: Chickie Silver, Michele Dillingham, and Deb Bierman. I had to let these very important people in my life know what was happening. Each one a longtime friend or colleague, and each like calling a brother or sister—my family of choice I had grown over the years. I realized, perhaps for the first time, the true value of those friendships.

My life is busy and often complicated, involving travel all over the country to make appearances, and I couldn't do it without the support of Emir Pehilj, who has been doing my hair and makeup for many years and has become a very dear friend and traveling companion. Emir stood ready to help me feel like myself from the very start, which went far beyond drawing eyebrows on me, or gluing eyelashes on after I lost them, or styling my wigs day in and day out. His emotional support was incredible. He actually shaved his head in solidarity after I shaved mine (he looked incredibly handsome). While I was going through my journey, Emir's sister was diagnosed with breast cancer; he knew all too well what was in store for her.

And then there's my incredible staff at our Joan Lunden Production offices in Armonk, New York. While my daughter Lindsay stands at the helm of my company, interfacing with all of the different corporations and networks that I work with, I rely each and every day on Elaine Capillo, my personal assistant, to be in constant contact with all the agents who book me for these events, to keep my schedule straight, and to handle the plethora of travel and event details that go into each appearance.

Then there is the "incoming," as I refer to it in the office, all of the daily messages that come in to me via the joanlunden.com website, Facebook, Twitter, and LinkedIn. Ali Barrella makes it possible for me to stay in touch with everyone. She works closely with Lindsay and Elaine and, more recently, Sarah, to keep our website up to date with blogs and postings from those on my staff and all of our wonderful contributors. It's a massive job, and it boggles my mind that my small but mighty staff can actually do it all. I applaud them, and I am so thankful for everything they do, especially during this past year, for they have all been on the breast cancer front lines right along with me.

There is no doubt that I have had the great luxury of the very best medical team at my fingertips. From the first day I started as cohost of *Good Morning America*, the best of the best have sat across from me, updating all of us with the latest and greatest information and research available in whatever was happening in their field. I'd be lying if I didn't admit a girl can get spoiled with that kind of exposure over the years. There wasn't any specialist I couldn't get to with just a few phone calls. When it comes to my personal health, I've always had an advantage in knowing I was somewhat protected and surrounded by what I considered to be my medical A team.

As a result, when I was diagnosed with breast cancer, I was able to get to the top doctors, seek out the best advice, and use my contacts in order to make the most informed decisions. Believe me, I recognize that many women don't have that option, and that is when and why I knew I had to write this book. I must thank my longtime physician Dr. Albert Knapp for helping me put together my top-notch oncology team. I greatly admire each one of the oncologists with whom I worked: Dr. Ruth Oratz, Dr. Tracey Weisberg, and Dr. Dickerman Hollister. I will be forever grateful to my wonderful breast cancer surgeon, Dr. Barbara Ward; my radiation oncologist, Dr. Ashwastha Narayana; and Dr. Gail Calamari, who first found my tumor. And finally, thanks to Beth Bielat, my physical and spiritual trainer, and my nutritionist, Dr. Robert Zembroski, without whom I might not have had the same journey, let alone outcome.

Contents

Had I Known

Introduction:
A Daunting Diagnosis

The truth will set you free . . . but first it will piss you off.

GLORIA STEINEM

Feminist, journalist, social and political activist, diagnosed with breast cancer in 1986

What if . . . *it's positive?*
       What if . . . *it's really bad news?*
What if . . . *it's worse than I think?*
Every horrible possibility was running through my mind as I quietly sat next to my husband in the radiologist's office that humid June afternoon. We were anxiously awaiting the results of my breast biopsy.

Wait, maybe it's nothing.

It can't be anything to worry about.

It's probably no big deal.

But what if . . .

Jeff and I held hands, our fingers tightly wrapped around each other's. I could feel sticky sweat between our palms.

Odd, I thought. Jeff's hands are rarely sweaty.

Still we didn't speak.

Not a word.

It was strange because, inside my head, it was as if someone were shouting: "You'll be fine. EVERYTHING WILL BE FINE!"

What will people say?

What if . . .

I'm sick . . . really sick?

What if the doctor walks through that door and says . . .

Just then, Dr. Gail Calamari opened the door to her office and entered the room. I could tell by her demeanor that it was not good news. It made me think of what they say about juries returning to the courtroom after deliberating and reaching a verdict: If you are guilty, they won't look at you.

Dr. Calamari did not make eye contact. She took her seat across the desk from us and slowly yet calmly began to explain that I had a 2.3-centimeter virulent, fast-growing breast cancer tumor that was very close to my chest wall. She said that I would need to see a breast surgeon as soon as possible. She recommended Dr. Barbara Ward, a terrific breast cancer surgeon who had a practice right in my area.

I did my best to stay stoic and numb. In my mind, that meant I couldn't cry. I absolutely wouldn't make eye contact with Jeff, because I knew that would break me. I've always had a need to feel like I'm in control and strong in situations like this. Don't get me wrong. I had that lump in my throat. I fought hard to swallow it back. I didn't want Dr. Calamari or Jeff to know it was there.

Whether I pulled it off?

Who knows.

But I didn't shed a tear.

Not then—not there.

Dr. Calamari knew what we had to do next. She called Dr. Ward for us and got me an appointment three days later, on Thursday.

Three days?

That seemed like such a long time to wait . . . an eternity, actually. However, we were told that Dr. Ward was highly sought after and worth the wait. I didn't want to appear ungrateful, so I said thank you and took the appointment.

I was still numb from the news when Jeff and I got into the car to head home. I think we were both in shock.

I could tell that Jeff felt really conflicted. I knew he wanted to be there for me, but he also needed to get to Maine. My diagnosis couldn't have come at a worse time of year for him. Jeff usually leaves for Maine right after Memorial Day weekend to make sure both of the summer camps he owns—Takajo for boys and Tripp Lake for girls—are in tip-top shape, his counselors are well trained, and everything is ready for the arriving campers at the end of June.

Let's be real. Is there ever a good time to be told you have breast cancer?

Still, all of his counselors were arriving that week, and Jeff needed to be there for orientation and training. There is always so much to do in the weeks before the start of the camp season. I didn't want to burden Jeff or preoccupy him with my health issue. I wanted him to put his focus on his campers and on opening for the summer. Those kids needed him more than I did right then. Trying to be the supportive wife, I assured Jeff that it would fine; Lindsay could accompany me to the appointment with Dr. Ward. Lindsay suggested that we have Jeff on speakerphone during the appointment so he could hear everything the doctor said at the same time I was hearing it.

We spent the rest of the day as if nothing had happened. Probably part of me was still in denial. There's also a part of me that believes you don't make payments on a debt not yet incurred. I wanted to believe the mass was small and innocuous. I had no reason to believe otherwise. I'm an optimist by nature. I have my mom to thank for that, good ol' glitzy Glady. That's what I called her, because she was like a shooting star—she could leave a room lit up long after she left.

"Hey, Joan Lunden . . . you've just been told you have cancer . . . what are you going to do next?" While I wished I were going to Disney World, the following day, we would have a house full of family celebrating the eleventh birthday of our oldest set of twins, Max and Kate, and their upcoming graduation from grammar school, so we wanted to be upbeat for them.

Deep down, part of me was contemplating not telling either set of twins about my cancer. Somehow I had myself convinced that I could get through the whole ordeal undercover. Maybe I could go stealth—you know, fly under the radar without anyone finding out.

Was I being naive?

Probably.

Was it wishful thinking?

For sure.

What mother wants to sit down with her children and utter the words "Mommy's sick"?

Let alone have to say, "I've got cancer."

"Cancer" is a super-scary word for kids.

Hello, it's a scary word for me.

I had no idea what I was dealing with yet. I didn't want to jump the gun or panic anyone more than necessary. Besides, I didn't want anything to detract from Kate and Max's big week.

On one hand, life seemed *so* normal as we set out to celebrate the kids' birthday, and yet it suddenly felt unfamiliar and strange.

One moment everything can be so incredibly wonderful—you tra-la-la through life, and then BAM!

All at once it feels as if you've been hit over the head with a frying pan.

That's what happens when you hear those words "You have cancer."

Jeff wanted to stay with me as long as he could, and he wanted to attend Kate and Max's graduation. When he left for Maine early on Thursday morning, I could see he was still reluctant, but there was nothing else anyone could do now except wait.

For now, we decided not to tell anyone what was going on except my oldest daughters. We really had no choice about that, as they would be by my side during the upcoming appointments. As far as I was concerned, the fewer people who knew, the better. If this diagnosis turned out to be nothing, I'd have fewer people to explain it all to.

Thursday morning arrived none too soon. Lindsay and I left the house and made the short drive to the office of Dr. Ward. On the car ride over, I was silently negotiating with God, as one sometimes does in these types of situations. *Just let me need a quick in-and-out surgery—perhaps a simple lumpectomy. And please don't say I need four weeks of radiation treatment. I heard that's every day of the week; how would I ever fit it into my demanding schedule?*

Somehow I convinced myself that the need for chemotherapy was remote, so I didn't even go there.

In fact, none of us was going there. Being a family of optimists, we all thought, *This happens to millions of women, it's not going to be that big a deal, it will be okay.* We all downplayed my diagnosis in the beginning.

As we drove up to the Sherman and Gloria H. Cohen Pavilion, a big impressive brick building across from the Greenwich Hospital, I couldn't help but think about how many times I had driven by that building, thankful I never had to pull in.

Well, my luck had finally run out.

As we made the left turn, I could feel my stomach get queasy. I was surprisingly nervous. I was hoping

Lindsay couldn't tell how uneasy I was. Once we walked into the doctor's office, Lindsay was doing her best to keep things light, talking about how nice the nurses seemed and how beautiful the office was. Lindsay was six months pregnant, so I was worried about her being stressed out, and I didn't want her being exposed to anything that could harm her or the baby.

A nurse immediately took us into a small room away from the regular waiting room where I could have some privacy. I was very appreciative of her sensitivity because I was desperately afraid of being seen. God forbid this news got out before I had the chance to tell the rest of my family, especially the rest of my children. That would have been devastating for everyone.

I liked Dr. Ward the minute she entered the room. She got right to the point, showing me the films of my breast and pointing out that my main tumor was not incredibly large, at only 2.3 centimeters.

"That puts it at Stage Two," she said.

Wait a minute. Did she say the *main* tumor? *There's more than one?*

That was new information.

Dr. Ward then pointed out a smaller tumor that was about one inch in front of the larger one. She called it a ductal carcinoma in situ (DCIS). While it was small, it nevertheless would also need to be removed.

Lucky me! I got a twofer!

Then she took a deep breath.

That is never a good sign from a doctor.

Apparently, the doctors had also found a spot on my liver during one of the tests.

"Probably nothing," she said, "likely just a hemangioma," but it would need to be checked as well. Dr. Ward was very matter-of-fact.

My first thought was: *More? I have cancer in other places?*

Breast cancer was one thing, but now she was saying I had something else to worry about!

She didn't skip a beat and kept right on with the rest of her explanation.

I will admit, I was quite stunned by the news of a second tumor, let alone a suspicious spot on my liver. It took me a minute to catch up with her. I am not sure I heard everything she was saying. At that point, she sounded more like Charlie Brown's teacher talking: "Wah, wah, wah." Thankfully, Lindsay was there with me, taking meticulous notes, and Jeff was on speakerphone.

Dr. Ward explained that the size of the tumor is only one consideration when assessing treatment. The pathology report that had come back on the biopsy was what would actually dictate my treatment. The pathology showed that my tumor was not estrogen-fed, like most tumors.

Okay, so hopefully that means that I didn't cause the cancer myself, taking all those estrogen supplements during menopause!

Yes, that is exactly what I was thinking!

Apparently, I tested positive for triple negative breast cancer.

WHOA.

That's good, right?

I tested negative for three things. That's *got* to be good!

But then the doctor explained that it wasn't good. What it meant was that I had a rare type of breast cancer untreatable by any of the typical targeted therapies and that my treatment would require months of chemotherapy.

For a moment, I thought I was going to puke.

Did she just say chemo?

I felt the blood drain from my head.

I wanted to appear calm and unfazed, but I'm afraid I failed at that attempt.

Lindsay could see the look on my face: I was frightened.

It was as if someone had sucked all of the energy and oxygen out of the room.

I couldn't breathe.

I needed air.

I wanted to get up and run.

But I didn't.

I just sat there. Motionless, staring at Dr. Ward. I was trying to stay with her so everything could remain clinical and academic. I wanted to hear everything she was saying. But honestly, I don't think I could process a word.

My first thought?

Will I lose my hair?

I asked Dr. Ward.

"Yes . . ."

Tears began to well up in my eyes, but I didn't cry. I wouldn't give cancer the courtesy of allowing the tears to fall down my cheeks.

Jeff had been patched in for the entire conversation. He was quiet, too.

Had he known this was what we were going to hear today, Jeff never would have gone back to Maine. He is the kind of husband who would have been by my side. I knew it was more painful for him to be so far away than it was for me. He understood me well enough to know I was putting up a strong front, but I am sure he sensed my fear and anxiety.

Still, there was just silence.

It took us all a minute or two to gain some composure.

And then the questions began.

Dr. Ward calmly explained that the type of cancer I had, triple negative breast cancer, was aggressive and fast-growing and would require that I go through a couple rounds of chemotherapy. The NCCN guidelines recommend adjuvant chemotherapy for all triple negative breast cancers that are over 0.5 cm in maximal dimension. The size of the tumor and the biomarkers make this the recommendation. Women with very tiny triple negative tumors sometimes escape chemo.

My first thoughts were: *Are you sure? Always? For everyone?*

She said, "Yes, this is the standard treatment for almost everyone with your type of cancer."

My type of cancer!

She said that many oncologists would have me follow the standard course of treatment that had been recommended for years: first a lumpectomy, followed by an ACT regimen of chemo (most frequently referred to as dose dense chemotherapy or ddAC-T), and then a round of radiation. A-C-T stands for the combining of three drugs; A-Adriamycin, C-Cytoxan, T-Taxol. When combining drugs, the goal is to enhance the treatment effect and not enhance side effects.

The order of these therapies might vary, depending on which oncologist I chose to work with. Chemotherapy given first is called neoadjuvant chemotherapy. By strictest guidelines, the purpose of chemo first is to shrink a tumor down so that a woman whose tumor is perhaps too big for breast conservation (lumpectomy and RT) could have breast conservation instead of a mastectomy after all. There is now some evolving data that suggest that if a woman gets chemo first, and has what is called a "pathologic complete response," that this will be a marker of a better cure rate or survival rate. This would be obviously the more powerful reason to give chemo first, but current clinical trials have not unequivocally proven this hypothesis to be true.

I was stunned and I was not quite prepared to hear all of this information yet.

Dr. Ward said that some oncologists were reversing the order of treatment and giving chemotherapy first, followed by surgery and then radiation. She explained that by approaching my treatment in this way, we would have the advantage of seeing whether the chemotherapy was shrinking the tumor, and when it came time for the surgery, there might be no tumor left, only the little metal markers inserted during the breast biopsy so the surgeon would see where it was when they were going in to remove it.

This way they could test any remaining tumor to see the pathology on it and be able to check their success based on evidence. If we took the tumor out prior to the chemotherapy treatment, there would be nothing to compare.

It made sense to me . . . I guess.

Dr. Ward said I would need to see an oncologist to find out what that treatment should be, but there were new drugs that were extremely effective, and I should end up cancer-free.

The information was coming at me so fast, I could barely keep up. I felt as if I had just been shot out of a cannon at supersonic speed into the world of oncology. There were so many new terms. It was an avalanche of information. I didn't want to be buried under this mountain of data I couldn't comprehend. I had no knowledge of my disease or treatment. In an effort to dig my way out, I started assembling my search and rescue team.

No one knew my medical history better than my primary care physician, Dr. Albert Knapp. He had been my doctor for over twenty years. I asked Dr. Ward if we could call and bring him up to speed and seek his advice. The two doctors compared their recommendations and then gave me the names of three oncologists at three different institutions. Feeling confused, I pushed them to tell me what to do, where to go, and whom to see first. It took some asking—in fact, a couple of times—to get more information on where to start, but Dr. Knapp recommended that I see Dr. Ruth Oratz, an oncologist affiliated with NYU Langone Medical Center, the following day.

Chapter 1
A Normal Doctor Visit That Wasn't So Normal

The only thing to really be afraid of is if you don't go get your mammograms, because there's some part of you that doesn't want to know . . . that's the thing that's going to trip you up. That's the thing that's going to have a really bad endgame.

CYNTHIA NIXON

Actress, diagnosed with breast cancer in 2002

June 5 started out like any other normal day. From the moment I awoke, the bright morning sun that peeked through my bedroom windows was a pleasant reminder that summertime had arrived. I love the change of seasons. Except for my pesky allergies, I embrace the loveliness that comes with the dawn at each time of year. Summer is especially welcome because it is a time for free-spirited play, bright sunny days, and beautiful warm breezy nights. It's the one time of year I look forward to slowing down a bit and taking a more leisurely approach to my otherwise hectic and over-scheduled life.

But not just yet . . . I had lots of work to get done and a few more things to check off my perpetual to-do list before I could even contemplate relaxing. Summer is always a busy time of the year for my family as the school year winds down and Jeff gears up for the camp season.

As for our four little campers, Max, Kate, Kim, and Jack, we'd already packed up their duffels and shipped those off to Maine, so at least I felt ahead of the game there.

My overpacked schedule had been especially challenged lately, making last-minute plans for Max and Kate's birthday party for their friends, then a family gathering the following day. As if that weren't enough, I'd been planning a small family dinner to celebrate their transition into middle school next year. (Middle school! I can hardly believe it. Where does the time go?)

My office is only fifteen minutes away from my home, but I wanted to get there early that day because

I knew I would be leaving for the two-thirty appointment to have my annual mammogram.

I have always been quite diligent about getting my breast screenings, and I now get the 3D mammogram followed by an ultrasound because I have dense and fibrous breast tissue, and this technology is far better than a standard mammogram alone. For many years, I didn't realize my breasts were so dense with fibrous tissue that, regardless of the technology, without an ultrasound, it was virtually impossible for the technicians to ever see a tumor growing in them. For years, no one ever explained that to me.

I had a fortuitous exchange in 2009 with Dr. Susan Love, one of the founding mothers of the breast advocacy movement. That conversation jolted me into reality on the need for women with dense breast tissue to have an ultrasound as well as a mammogram. It likely saved my life.

Now, to be fair, I had heard this advice once before. The first time was during an interview with Dr. Judith Reichman for a DVD I was hosting for Eli Lilly on type 2 diabetes management. Dr. Reichman is considered one of the leading voices in America on women's health issues.

During a break, I took advantage of some one-on-one quality time with this esteemed expert on women's health. At the time, I was contemplating hormone replacement therapy (HRT) as a treatment for menopause and was considering estrogen, progesterone, and testosterone supplements. I told Dr. Reichman that I had no family history of breast cancer but that I had always wondered if it was really okay to be on HRT. I did have a family history of heart disease and had been told that research showed HRT provided some cardiac protection.

Dr. Reichman said that every patient is different. While HRT may eliminate many of the effects of menopause, there were still concerns among the mainstream medical community, especially when you combined estrogen and progestin, which is a synthetic progesterone. She was also quick to ask if I got regular mammograms. I assured her that I was diligent about getting my mammograms, but my doctor always seemed to ask for more pictures, saying something about my breast tissue being dense and fibrous.

With this information, Dr. Reichman advised that I should always follow my mammogram with an ultrasound. Although her advice was crucial for someone like me with dense breast tissue, unfortunately, this is not universally accepted as the standard of care in the United States. As a result, some women who request an ultrasound may find themselves paying out of pocket for their test because insurance will not cover the cost. Why? For some, an ultrasound is usually looked upon as a problem-solving test—not a lifesaving test.

Despite her advice, I'll admit, I didn't ask my doctor to add the ultrasound that year, or even the year after. I don't even have a good reason or excuse. I just didn't do it. However, a few years later, I was given that advice again while interviewing Dr. Susan Love for *Health Corner*, a show I was hosting for Lifetime. The interview was about breast cancer and what age women should start having mammograms.

During the interview, Dr. Love spoke about a new study she was involved in to shed some light on what causes breast cancer. She said that the majority of women who get breast cancer have *none* of the known clinical risk factors.

Off camera, Dr. Love asked if I was getting regular mammograms. I told her I was and that it was incredibly nerve-wracking because I always had to go back in for more images due to my dense breasts.

That was when she said that if I had dense breast tissue, the mammogram might not be able to pick up a cancerous tumor. I most likely needed an ultrasound every year, in addition to my mammogram, to be thorough in my examination.

Okay. Wow.

That was the second time I'd heard this advice.

This time I'd be a fool to ignore it, that's for sure.

At my next mammogram appointment, I asked my gynecologist to write a prescription for my annual mammogram and an ultrasound. It became a routine part of my yearly exam, and until this appointment, I'd always exited my radiologist's office and gone on my merry way.

While I know there are some women who are afraid to get their boobs squeezed, I'm not one of them. I didn't really enjoy it, but subconsciously, I also didn't believe I could ever be one of the statistics—you know, one of those one-in-eight women. Perhaps it was the power of positive thinking that kept me in that mindset every year. Or, more likely, the assumption that I was risk-free.

When I arrived at the radiologist's office, the nurse showed me to a tiny closet of a changing room, handed me the powder-pink robe, reminded me to wipe off any deodorant I was wearing, and said, "We will come get you shortly."

I quickly changed, took my place in the waiting area with half a dozen other women, and started making a mental checklist of where I was headed as soon as I was done here. I looked around the waiting area and noticed some of the women were flipping through magazines—not reading, just nervously flipping the pages. Others had their heads angled straight down, looking at their phones, furiously swiping up with one finger, or reading their Kindles.

No one was making eye contact. Women don't seem to make small talk in this area. I've always thought it was because we were all nervous or feeling awkward, waiting half naked in our matching robes. I couldn't help thinking, *In any other social setting, it would be a faux pas to be wearing the same outfit as the woman next to you. Here, it's all the rage.*

What would Anna Wintour think?

Does she fashionably belt her robe?

Maybe she ties hers with an Hermès scarf?

Oh, what we do and think to pass the time waiting for our mammograms!

Even though I didn't consider myself at risk for breast cancer, deep down, the process made me feel a little fidgety and apprehensive. Sitting there waiting was a bit nerve-wracking. I think it's safe to say that every woman in that room, anticipating her name to be called, secretly worried she may not escape her visit without hearing those dreaded words: "We saw something that concerns us."

Over the years I have been called back for further images several times. Even with the extra look, I have always been able to walk out the door with a clean bill of health, let out a giant exhale, and start breathing once again.

Am I alone in this? Don't we all sit there, holding our breath, waiting to hear the words "You're all done, and you are clear for another year"?

Whew!

We can't get dressed and get out of that place fast enough.

Once I was done with the mammogram, I still had to wait for an ultrasound room to open up. While this took a little more time, it was always worth it for my peace of mind. Besides, I was breathing a lot easier: The worst part was behind me. The ultrasound is completely painless, merely a precautionary step. I considered it an extra insurance policy. Ever since I started getting ultrasounds, there had been no sign of anything to be worried about. I was practically home free.

As my name was called and I walked down the hall, my mind was already past the possibility of something going wrong and on to birthday cake and decorations. After all, I did just get a clean bill of health across the hall, right?

Another year, another clean mammogram, I thought.

Exhale.

Smile.

All I had to do was get naked one more time for another technician to poke around at my boobs!

It is so weird and embarrassing.

And yet it is a lifesaving exam, so we just have to get over feeling uneasy and uncomfortable with allowing a perfect stranger to fondle us in all sorts of awkward ways. "Hiya, nice to meet you . . . Yup, that's my boob!"

To break the ice, I always engage with the technician as soon as she comes into the exam room. I chitchat and joke around as my way of coping with the discomfort I'm feeling with the situation. Somehow, talking with the technician about the latest news story or the hottest TV show distracts me from how embarrassing it is to have a stranger staring at my half-naked body.I know I'm not alone here. They must have all sorts of funny, nervous, and sometimes odd conversations, because they go through this fifty times a day. Without missing a beat, they usually play along with my banter as I lie on the exam table with my right arm extended over my head and my right breast totally exposed.

The tech squeezed some of the cold gel all over my breast; the gel would somehow enhance the image.

(I've always thought they should keep that goo in a warmer—kind of like the warmers you can get at Babies "R" Us to keep your baby's bottle warm—so that it isn't so uncomfortably cold when they spread it across your bare breast. Brrrrrrrrr! Doesn't that seem like an obvious idea?)

The technician then guided the cold ultrasound instrument in a circular motion around my right breast, as she always did. However, this time she kept going back to one spot, again and again. It was right around this time that she stopped engaging in dialogue with me.

She checked my left breast but said she wanted to go back once more to check my right breast.

Uh-oh.

I remember thinking, *That can't be good. She must be seeing something bad.*

Techs aren't supposed to tell you *anything*, good or bad, but the journalist in me felt compelled to ask her anyway.

Of course, she said she wasn't allowed to tell me anything. However, she did say she needed the radiologist, Dr. Calamari, to come in and take a look at something.

Shit.

She left the room while I lay there.

My right arm was still up over my head, the clear cold gel smeared all over my right breast.

I suddenly felt frozen in time.

It seemed like hours passed before the radiologist came in, though it was probably more like four or five minutes. Our perception of time can be so warped when we are waiting for the worst.

Dr. Calamari looked closely at the screen and then took the ultrasound instrument and moved it back and forth, again and again, over the same spot on my breast. Without saying a single word, she snapped more images, a lot more. Then she said, "Get dressed and come see me in my office so we can discuss this."

I didn't know exactly what she was about to say, but I did know I didn't like the sound of it.

My heart was racing almost as quickly as my thoughts. I had no time for this. I had things to do. I had parties to plan, family coming for the weekend, kids leaving for camp, my husband about to leave for the summer, work obligations, commitments . . . I was making an endless list of all the reasons Dr. Calamari's news simply wouldn't fit in to my schedule right now.

No.

Absolutely not.

Not now.

Not ever.

No.

Hell no.

No, no, no, no, no, no, no, no, no, no,

PLEASE
GOD
NO.

YES. Dr. Calamari told me she had seen something on the screen that looked like it could be a tumor at the back of my breast, near my chest wall. She needed to do a needle biopsy to see exactly what we were dealing with. She said she would like to do it right there and then, that afternoon.

What?

Really?

Right now? You want me to go get undressed again and let you stick a big needle in my breast and cut a small piece to send to a lab?

WOW!

NO!

These were just the thoughts I was having—I knew I was going to have the biopsy—but I definitely wasn't prepared to do it that afternoon. I had the perfect out, and I shared it from my best "Mother of the Year" point of view: "You see, I promised my nine-year-old daughter, Kimberly, that I would be at her gymnastics show this afternoon, and I don't want to disappoint her."

"Oh, I completely understand, my daughter used to take gymnastics at the same place when she was younger. I remember those end-of-the-season shows for parents so fondly. You don't want to miss that. You can come back in the morning and we'll do it then," Dr. Calamari said.

Whew.

At least I bought a little time.

Dr. Calamari explained that she was usually reluctant to allow a woman to leave the office without doing a biopsy on the spot, because women all too often never return, fearful of potentially bad news. She also pointed out that the following day was a Friday and that normally she didn't do biopsies on Fridays, since her patients had to wait the weekend to get the result, and women find that excruciatingly stressful.

Under the circumstances, we agreed that I would come back first thing in the morning for the biopsy and would return for an early appointment on Monday to hear the results.

Dr. Calamari reminded me that while it looked like a cancerous tumor, it might turn out to be just a cyst. If that was the case, once the needle pierced the cyst, the fluid inside would be extracted with the biopsy needle, and life would go on as if nothing happened.

Life would go on as if nothing happened.

Those were the words I couldn't get out of my head.

Those were the words I needed to focus on.

Those were the words I desperately wanted and hoped and prayed would be true.

On the way home, I called Jeff to tell him what had happened. I really wanted to make light of it, so I wouldn't worry him and make him feel bad if he couldn't be there should something be wrong. After all, I didn't know what I was dealing with yet. There was no reason to panic him.

How do you make this kind of call?

"Hi, sweetheart, I may have cancer, but don't worry, I'll be home in time for dinner."

"Hi, it's me. Remember those breasts you liked so much? I've got good news and bad news . . ."

I tried to sound somewhat nonchalant during the call, but my emotion started to surface, and my voice began to break. All of a sudden a few tears were trickling down my cheeks. Fortunately, he couldn't see them. I always wanted to appear the healthy vibrant woman Jeff met and fell in love with, and I wanted to be strong for him, especially at this time of year.

I told Jeff not to worry, that it may turn out to be a simple cyst, and if so, it would drain right out when she inserted the needle.

"Really, it may be nothing at all, sweetheart, so we don't need to be worried quite yet." I did my best to sound convincing, though I don't know if I pulled it off.

I have an exceptionally close relationship with my three older girls. Now that they are grown up, they're like my best friends. But I knew this would be upsetting news. I thought long and hard about burdening them with it.

Jamie is my oldest. She lives and works in New York City as an executive in PR, a relentlessly demanding field. She loves her work and is really good at it, but I feel she is always stressed to the max from its demands. As the saying goes, she is truly "a chip off the old block." She is an overachiever, a loyal hard worker who will go to the end of the earth to be the best at what she does, whether it's jumping horses in an international competition or landing the biggest, most impressive client for her company. Jamie and I spent a lot of time together on the road when she was growing up because she was a nationally ranked athlete in the equestrian world. We ventured to top competitions all over the country and became very close along the way. I think we are similar in our drive and our passion, although I gave up jumping horses after my last broken shoulder. Now I jump big fences only in my dreams, but I'm really great in those dreams. Jamie is still really great in real life.

My middle older daughter, Lindsay, worked in the fashion industry after graduating from the University of Pennsylvania. One day she asked to go to dinner with Jeff and me—on a Saturday night! *Really? What big news could she want to share?* we wondered. That night Lindsay presented her plan to work with me at my company and help take it to new levels. Believe me, that is exactly what she has done. It is the most wonderful gift when your daughter joins you in your life mission and dedicates her days to your goals, helping women stay healthy and happy. We have become incredibly close since she came to work with me and have traveled the country together for years now. However, when I received my troublesome news, Lindsay was six months pregnant, so I was terribly worried about divulging it to her and having it possibly impact her or her unborn baby in a negative way.

Then there was Sarah, who was my "baby" until the two sets of twins were born. Actually, she still is my baby. Like Lindsay, she graduated from the University of Pennsylvania. Sarah immediately joined the page program at NBC. It is the ultimate training ground for anyone interested in working in television broadcasting and entertainment. Pages work as studio tour guides and audience ushers on NBC shows including *The Tonight Show* and *Saturday Night Live*. The competition is so stiff that only 212 pages are selected each year out of the nearly 16,000 applicants, making it a harder program to get into than Harvard! Luckily for Sarah, her dream of becoming a page panned out.

NBC executives hire from that pool of workers and she was soon tapped to be the personal assistant to Frances Berwick, the president of Bravo. It was such a wonderful learning experience for Sarah, but she decided to pursue the other side of the business—the production side—and moved to Hollywood to work for a production company.

While she lived the farthest away from me, Sarah was the one I needed to connect with first. You see, I know my children. I inherently understood that she would feel bad because she wouldn't be able to come right over, wrap her arms around me, and tell me that everything would be all right. That's the kind of compassionate girl Sarah is.

So after I spoke to Jeff, I called Sarah and shared my news. I wanted her to hear it from me. I told her that it might be nothing, but I was having a biopsy in a few days. I would let her know any news as soon as I had something to report. I did my best not to sound panicked or scared. I just stated the information as I knew it.

At first Sarah thought I had a "different" kind of cancer. It couldn't possibly be the kind of cancer that required chemo or killed people. She thought perhaps I'd had a bad mammogram and ultrasound, and they were just being precautionary. She told me that we'd take everything one step at time, sounding completely calm and rational, because the reality of what I had shared hadn't set in with her yet. And how could it? She was three thousand miles away—and frankly, I'm not sure the reality had started creeping in for me yet, either.

Later that day, Jeff and I left the house as if nothing were wrong. As far as I was concerned, there wasn't anything wrong. Out of mind, out of possibility! At least that's what I kept trying to convince myself.

We took Kimberly to her gymnastics demonstration, sat behind the glass partition, and watched with the other parents as she popped on and off the balance beam and the uneven bars. I smiled and took pictures, but despite my best efforts to ignore the ten-thousand-pound elephant, I couldn't erase the obvious from my brain for a second.

Chapter 2
The Big C

Courage is being scared to death and saddling up anyway.

JOHN WAYNE

Actor, diagnosed with lung cancer in 1964;
credited with coining the term "The Big C"

When I awoke the morning of the biopsy, I did my best to convince myself there was nothing to worry about. I didn't have time for cancer. There was no solid evidence of it, anyway.

I needed proof.

Some concrete data before panicking.

You know what?

I needed to stay calm, center my thoughts, and breathe.

After all, the doctor had said it could turn out to be a cyst, and she would know better than anyone.

Right?

I mean, what else could it be?

It couldn't be a tumor.

No way.

I raced around, keeping myself nervously busy until it was time to face the possibility of some other reality.

Maybe I should just call and postpone!

No, you can't do that! That's not you.

What's the point in postponing possible bad news?

Damn. My conscience got the better of me. I'd promised Dr. Calamari I'd be there first thing that morning, and so I faced the music.

Dr. Calamari spoke to me as she was doing the ultrasound-guided aspiration and core biopsy. I was really nervous about the procedure.

Why?

Because it involved the doctor sticking a needle in my breast and poking it around until she found the suspect mass, then clipping little parts off of it and pulling them back out.

Ouch!

That sounded like it would hurt.

A lot. I was lying facedown on a table that had an open area where my breast fell down through. Before the procedure began, I was given several shots of a local anesthetic to numb the skin and breast tissue around the suspicious area. Then an MRI machine was moved into position to take images of the area in question. The radiologist watched the MRI image of the breast (an imaging technique that generates detailed 3-D pictures) on a computer screen in order to guide the positioning of the hollow needle into the breast until it reached the mass. Once the needle penetrated the mass, the doctor would know immediately whether it was a fluid-filled cyst or not. If it were a cyst, the fluid would be extracted with the needle, which was what we were hoping for. Otherwise, it was a tumor.

The room became suddenly and oddly quiet. You could have heard a pin drop when Dr. Calamari performed the procedure, searching and probing around the inside of my breast.

Unfortunately, it wasn't a cyst.

Dr. Calamari calmly and softly said she had found a solid mass and was taking a number of snips to send to the pathology lab. She then informed me that she'd inserted a tiny metal clip, which would be used later, during a lumpectomy, to find and remove the *tumor.*

There it was . . .

That word . . .

The *one* I didn't want to hear.

TUMOR!

I had a cancerous tumor inside me.

The doctor told me to get dressed and meet her in her office. I was lying there, still facedown, much as you do when you're on a massage table, but this was anything but relaxing. My whole body was extremely tense. Being told I had a tumor growing inside my body felt like I was having an out-of-body experience. It was an insanely helpless feeling.

The procedure hadn't been as bad as I had imagined, but the result was so much worse than I ever could have dreamed.

Welcome to my nightmare.

I pulled myself up from the table, walked silently to the tiny dressing room, and mindlessly put my clothes back on.

I didn't want to feel panicked . . . yet . . . so I chose to feel *nothing*.

I certainly didn't want to think about the possibility of dying.

No way was I going there.

Jeff, who had come back from Maine for this procedure, was already siting in the office when I walked in. I don't know if he could see the terror on my face or not. Dr. Calamari explained that the biopsy would tell us a lot more, but there was no doubt that I had breast cancer.

Dammit, the proof was there, and there was no denying it.

Time stood still for a moment.

Some things in life are so surreal that your mind can't process them easily.

This was one of those moments.

Now, I'm not the kind of woman to break down and cry. There have been occasions in my life when I thought maybe I'd be better off if I were that type—when I wished that I could just let go, be more vulnerable and sob, even a little. It's exhausting always feeling the need to put up the strong front.

However, this news went against everything I thought about myself.

I was a healthy vibrant person!

Wasn't I?

I felt instant guilt that I was putting Jeff in this situation.

I needed to be strong and healthy for him and for all of my children.

I wanted time to stay frozen, but the cruel reality was about to be laid out by my doctor.

I wasn't ready to hear it.

And yet time stands still for no one.

Once the pathology report came back, we would know the best course of action to take, but it would most likely require surgery and four weeks of radiation. There was always the possibility that I'd need chemotherapy, but it was too early to make that call, especially without the test results.

Since it was Friday, I would have to wait the weekend to see where it would all land. Dr. Calamari had warned me that this was why she didn't like to do these tests on Friday. But what choice did I have? I was in it now—deep.

"Go about your weekend like everything is normal," she said.

But there was nothing normal about my weekend; I simply couldn't get the thought out of my mind that I was facing *c-a-n-c-e-r.* I'd capitalize the letters here, but I wasn't ready to give the disease that much power in my world—not then, not now, not ever.

The thing about cancer is it doesn't care about your plans. No matter how uptight and frantic I might have been on the inside, I needed to put on a brave face and pretend nothing was different, because we were hosting a couple dozen kids for Max and Kate's birthday party the following day.

HAD I KNOWN • 39

Yup, we took several carloads of kids to a place called Bounce to celebrate Max and Kate's eleventh birthday. This popular party place has more kinds of trampolines than you could ever imagine, and the kids jump and jump and jump around nonstop for an hour and a half and then end it all with cheese pizza and birthday cake. Seriously, just watching, up and down, up and down, up and down, is enough to make you want to puke. But *they* all love it, and that's what counts, right?

It was a time for wide-eyed smiles, candid photos, and celebrating. I looked at my children and thought, *They're so young and innocent. How will I ever tell them?*

I could feel a lump beginning to swell in my throat as I pondered, *What if it's really bad—like, terminal— and I'm not around as they celebrate these birthdays in years to come?*

Can't go there.

Don't go there.

STOP IT!

Dammit.

I don't want to be sick.

I don't have time for cancer.

Then I quickly realized I needed to get my head back into the party.

I needed to be there for my kids.

On Sunday we had another celebration planned: a combo birthday party for the kids and baby shower for Lindsay, who was expecting in early September. I'd been planning a midsummer baby shower for a larger group of women, but just about everyone in our immediate family would be in Maine on that date, so we took this opportunity to celebrate. Everyone there was family, and yet no one in the room knew anything about the stark reality of my cancer except Jeff and me. As soon as I began telling people, it would instantly become more real.

I didn't want that.

And I didn't want any pity.

And I surely didn't want to think of myself as a sick cancer patient.

I wanted to remain quiet as long as I possibly could.

But I felt like I was living in two parallel universes. One was happy-go-lucky, and the other was super-dark and scary, with potential for the worst-case scenario: early death.

And boy, did I know something about sudden early death.

My dad was an avid pilot; we flew a lot as a family, and we often accompanied him around the country as he spoke at medical meetings and sometimes assisted other doctors in difficult cancer surgeries. Our home in

Northern California was literally built around an airplane hangar, with a taxi strip out to the runway of a small fly-in community.

When I was thirteen years old, I watched my father take off on a short business trip to speak at a cancer convention in Southern California. He had asked my mom, my brother, and me to accompany him to the conference, since it was a brand-new plane. At first my mom said no, thinking that we shouldn't miss school. But at the last moment, she changed her mind and we drove home. As fate would have it, just as we pulled up to the house, my father's plane was lifting off the runway. We missed him by moments. I stood and waved goodbye, totally unaware that this was the last time I would ever see my father. His plane crashed in Malibu Canyon as he was returning home from that medical convention with another cancer specialist.

But my dream to follow in his footsteps lived on.

Chapter 3
I Thought It Could Never Happen to Me

I had a false sense of security about cancer. It won't happen to me. Well, you know what? It did! I feared for my life, and then for my career, but I learned that it helps to turn fear into action.

DIAHANN CARROLL

Actress, diagnosed in 1997 with breast cancer

Accepting that you have a potentially deadly, fast-growing type of cancer in your body is nothing short of surreal.

How did this foreign mass, this tumor that was out to kill me and spreading its venom with each passing moment, get inside me in the first place?

Had I done something?

Taken something?

Eaten something that had brought this on?

I couldn't help but wonder how I'd brought this on myself.

While I didn't have time for self-loathing or self-pity, I definitely found myself looking back and wondering how it had happened to me. What actions did I have to take responsibility for that might have brought this on?

We've all heard the saying "Ignorance is bliss." Well, that was the only explanation or excuse that I could conjure up when trying to justify why I had always felt immunity from breast cancer. While I have been a broadcast journalist for over three decades, I am going to be completely candid and let you in on my totally lame rationales and the absurd reasons I believed I was never going to get breast cancer.

At the top of my list was the belief that since I am a health advocate for women and spend most of my time crisscrossing America to address crowds about staying healthy and ways to increase their longevity and avoid deadly diseases, I couldn't be affected by anything I was talking about.

No way!

I somehow convinced myself that since I was out on a speaking circuit, telling others how to avoid disease through exercise, nutrition, and making better lifestyle choices, I somehow got a free pass.

Don't even say it.

It was pretty bold and, yes, overconfident.

This next reason is for your personal amusement: I actually felt I was somewhat immune because my father had been a cancer surgeon. Okay, I get it; it's kind of like thinking if your father is Steve Jobs, your computer won't get a virus.

While I recognize that what my father did for a living didn't earn me a get-out-of-jail-free card, it often brought up the issue of my family medical history. For as long as I could remember, I had contended that there was no significant history of breast cancer in my immediate family; therefore, breast cancer happened to *other* women.

Logical, right?

In reflection, why did I think that?

Had I known that only 15 percent of women diagnosed with breast cancer have a family history of the disease, I probably wouldn't have been so quick to run with that stance. When I discovered this ASTONISHING statistic, I felt so left in the dark. How was it possible that, after years of being an educated health advocate, I had no idea that such a small percentage of women could use their family history as an indicator?

Something wasn't right! If I wasn't aware of that information, surely others were as in the dark as I was.

And the more I thought about it, was I really so certain that I didn't have any breast cancer in my family medical history? I hadn't heard of any, but now I was questioning whether I knew this for a fact. My family never lived close to any of our relatives, on either side of the family, so I hadn't spent much time with any of them and consequently hadn't been privy to casual family conversations between grandmothers and aunts. I was a child. I didn't keep up with who had what illnesses. Looking back, I can't say why I was so adamant that breast cancer didn't run in my extended family. I had no facts whatsoever to back that statement up. That's the first rule they teach in journalism school: Facts first, story second. And yet I continued to answer that question the same way, over and over.

It made things easier to rationalize when I decided to take hormone replacement supplements so I didn't have to suffer from hot flashes, sleepless nights, and all the other not-so-fun symptoms of menopause.

I call this the ultimate "inside job," because I was telling myself I had *nothing* to worry about. While I had heard there could be some elevated risk of developing breast cancer from taking hormone replacement during menopause, when I assessed the pros and cons, anything that made life more comfortable was fine by me, even if this justification had potentially dangerous consequences.

Wow!

Who was I fooling, anyway?

Who was I putting one over on so I didn't have to suffer hot flashes?

Yeah, I know hot flashes suck, but in retrospect, possibly dying sucks a lot more!

Some of my friends and family raised an eyebrow or two when I shared that I was taking hormone replacement therapy to minimize my symptoms. And while I've been told that taking HRT all those years may have had absolutely nothing to do with my developing breast cancer, unfortunately, once you're diagnosed with the disease, it's hard not to beat yourself up and wonder if every bad decision you've ever made along the way, from drinking diet cola to lying in the sun, has finally caught up with you.

So what are some of the other risk factors that I should have been concerned about?

Duh!

They were so basic that I essentially overlooked them.

For starters, *getting older* tops the list. While we all know it's inevitable, we never really anticipate the impact. Acording to the American Cancer Society, about one out of eight invasive breast cancers develop in women younger than forty-five. About two out of three invasive breast cancers are found in women fifty-five or older. In fact, the aging process is the biggest risk factor for breast cancer. That's because the longer we live, the more opportunities there are for genetic damage to occur in the body. And as we age, our bodies are less capable of repairing genetic damage. That's why it's so important to take care of ourselves by eating right, exercising several times a week, maintaining a healthy weight, not smoking, and living a healthy life-style as we age. I'd always thought I was doing this. Had I known that the choices I'd been making before I was diagnosed weren't all the right ones, I would have changed things a long time ago.

But more on that later.

Next is *dense breast tissue*. About 45 percent of women aged forty to seventy-four have dense breasts. The younger you are, the denser your breasts are. Dense breasts have more fibrous and glandular tissue than fatty tissue, which can make it difficult for a radiologist to detect cancer on a mammogram. Dense tissue, like cancer, shows up white on the mammo-gram. Fatty tissue is dark, so any cancer is more visible. While I was aware that I had dense breasts, and that made it difficult to spot and diagnose cancerous tumors without an ultrasound, what I didn't realize was that women with dense breasts have been shown to be four to six times more likely to develop breast cancer. Only age and the much-talked-about BRCA1 and BRCA2 mutations increase their risk more. And since I'm mentioning BRCA1 and BRCA2, these are human genes that produce tumor suppressor proteins. According to the National Cancer Institute, specific inherited mutations in BRCA1 and BRCA2 increase the risk of female breast and ovarian cancers, and they have been associated with increased risks of several additional types of cancer. Together, BRCA1 and BRCA2 mutations account for about 20 to 25 percent of *hereditary* breast cancers (1) and about 5 to 10 percent of *all* breast cancers. In addition, mutations in BRCA1 and BRCA2 account for around 15 percent of ovarian cancers overall. Breast and ovarian cancers associated with BRCA1 and BRCA2 mutations tend to develop at younger ages than their nonhereditary counterparts. A harmful BRCA1 or BRCA2 mutation can be inherited from a person's mother or father. Each child of a parent who carries a mutation in one of these genes has a 50 percent chance (or 1 chance in 2) of inheriting the mutation.

While experts say that *having children past the age of thirty* poses a slightly increased risk of developing breast cancer, I didn't really consider it to be a huge contributing factor, though I had my first child when I was thirty and two more before I turned forty. All three of those pregnancies were naturally conceived, making the risk very low. In the scheme of contributing factors, this wasn't one I could give a lot of weight to.

On the other hand, *extended exposure to estrogen and progesterone* was something I needed to consider as a possible factor. You see, like a lot of women of my generation, I had such terrible menstrual cramps as a teenager that I was given birth control pills to help regulate my periods. I continued taking them until I was in my late twenties and began trying to get pregnant. Throughout the course of my adult life, I took oral contraceptives until I reached menopause.

When I married Jeff, he and I went through two years of fertility treatments—several rounds of unsuccessful in vitro fertilization—pumping all sorts of hormones into my body, trying to conceive a baby together. When we weren't able to conceive, we turned to a surrogate to help us build our family, and I went directly on hormone replacement therapy.

Of course, I didn't look at this buildup as a problem over the years, but it probably had some kind of impact. I suppose I'll never know, and really, what's the point of trying to figure it out now anyway?

I'm going to let you in on a little secret.

Promise not to tell anyone?

In spite of my love of fitness, there were long periods in my life when I did absolutely no exercise at all.

Damn!

I never thought that one would come back to bite me in the butt—or should I say boob—in such a big way.

It turns out that *lack of regular exercise* can impact the likelihood of developing breast cancer. So the next time I'm lying in bed, deliberating whether or not I should get up and hit the gym, you can bet I'm going to move my ass!

Thankfully, I'm what most people refer to as "a cheap date."

Why?

I don't really like the taste of alcohol.

Other than a glass of good wine every now and then, I can live without booze, so *consuming alcohol,* which can be an indicator for developing breast cancer, wasn't a real issue for me.

And just when I thought I had escaped the wheel of cancer roulette, it's as if someone blurted out: "But wait, there's more!"

It's not like I bought a set of Ginsu knives.

(Gee, wouldn't that have been nice!)

As researchers learn more about how cancer cells develop, they are concluding that there are several other stressors on our immune system that may be important risk factors for developing cancers, including breast cancer.

"Stressors" really got my attention!

Why?

Because *lack of sleep* and *stress* are two of the biggest perpetrators science is studying that can and likely do contribute to a higher risk of developing breast cancer.

After spending nearly two decades working on *Good Morning America,* I understood the meaning of sleep deprivation. I'd gotten used to my routine of waking up at three-thirty A.M. so that I could leave my suburban home by four and arrive in New York City no later than five for hair and makeup at the television studio. Those were long days, with countless road trips and lots of pressure to perform at the top of my game.

Did I mention that I had three young daughters at home who needed me from the moment I walked through the door until the second they went to sleep?

My day job couldn't interfere with my responsibilities as a mom.

Who had time to sleep?

On a good night, I hoped for six solid hours of sleep, but that meant getting into bed by nine-thirty.

Yeah, that didn't happen very often.

I was supposed to be snug in my bed, lights out, eyes shut, when my daughters wanted homework help or were on the phone making plans with friends for the next day after school.

Okay, not to be an alarmist, but who doesn't have stress in their lives? However, those of us who live our private lives in the public eye sometimes have our stresses magnified!

Does that make my stress bigger or worse than anyone else's?

No.

But it sure makes it more public and sometimes better tabloid fodder!

When I look at this list of possible contributing factors, I see quite a few matches that I could place a check mark next to.

Does that mean I somehow gave myself breast cancer?

I don't know the answer to that.

I can tell you I worried about it.

A lot.

I certainly felt guilty for some of my decisions.

The only purpose of regret is to challenge decisions we've made along the way and to question the course of our lives. Any way you looked at it, regret wasn't going to change my diagnosis.

When I thought about my father's life's work, I realized he died with so much left to do in the fight against cancer. Somehow I understood that through my diagnosis, I was being given the chance to carry on his legacy. If I shared my story and went public with my battle, it occurred to me, I might have the opportunity to help save even more lives than my dad—who was taken much too soon.

For the past twenty-five years, much of my work had been dedicated to and focused on health and wellness. To be fair, I was the image of health without realizing that I wasn't the picture of it. With my breast cancer diagnosis, my mission in life had just grown. You see, every challenge we face in life is really just an obstacle waiting to be turned into an opportunity. It occurred to me that my purpose had now become carrying on my dad's work. I had never been given a more important assignment. I wanted to—no, make that needed to—inspire others to protect their health. It had to become a necessary commitment, one I would accept with dignity and grace. My focus would be primarily on breast cancer. I couldn't think of a better way to honor my father and the work he did.

I'll admit I haven't made perfect choices every single day—who among us has?

That's what makes each of us human.

One thing I know for sure: I can't dwell on what *was*.

I have to place my focus on what *is* so I can get to what *will be*.

Chapter 4
Choosing the Path
of Least Regret

The only person who can save you is you; that was going to be the thing that informed the rest of my life.

SHERYL CROW

Singer/songwriter, diagnosed with breast cancer in 2006

When the pathology report came back, it said that my breast cancer cells tested negative for estrogen receptors, progesterone receptors, and HER2 receptors. Testing negative for all three means that the cancer is *triple negative*. Triple negative results mean that the growth of the cancer is not supported by the hormones estrogen and progesterone, or by the presence of too many HER2 genes. Therefore, triple negative breast cancer (TNBC) does not respond to regular hormonal therapy like tamoxifen or therapies that target HER2 receptors , such as trastuzumab. As I would quickly discover, about 10 percent of all breast cancers diagnosed—about one out of every ten—are found to be triple negative.

And now, as a result of this diagnosis, I was facing the biggest battle of my life: *saving* my life.

Where does one even begin with this fight?

For a lot of women, this is the stage where they freeze in fear.

Boy, can I understand that.

It's scary as hell.

While I was terrified, I have never been the kind of woman who crumbles in distress; I use my anxiety to rise to the occasion, and I take all of that negative energy and channel it into something positive. In this case, the only way I could cope with the reality was to get on my surfboard, ride this unintentional wave, and yell, "Cowabunga!"

I inherently knew I would want to choose the best and most effective course of treatment available today—one that would kill the cancer that had taken up residence inside of me. My real angst was over finding my way to that answer.

It wasn't going to be an easy walk. You see, there are several different ways to attack my kind of breast cancer. I could speak with the three leading oncologists about which one was the best, and I would get three completely different answers. It wasn't going to be a black-or-white decision. In the end, it would be up to me to make the decision that felt right for me.

Once I was diagnosed with breast cancer, I felt as if everything happened so fast. I know it doesn't move at lightning speed for everyone, but my treatment was definitely on a fast track. The day after I met Dr. Ward, Lindsay met me at the office of Dr. Ruth Oratz, one of the leading oncologists in New York City. Dr. Oratz's treatment center catered to breast cancer patients, so there were only a few other women in the waiting room when we arrived. Dr. Oratz's office had a warm and fuzzy atmosphere that I found surprisingly welcoming. It wasn't a massive, institutionalized cancer center. It was a private practice that felt personal and was the antithesis of a large hospital setting. It was exactly what I needed at that moment. I hadn't told anyone about my diagnosis yet and was petrified someone would blow the whistle on me and take my news public before I was ready. Although I was still in something of a state of shock, I instantly felt safe the moment I walked through the door.

Dr. Oratz and I clicked from the moment we met. I wasn't surprised, since she is married to Dr. Albert Knapp, my primary care physician and one of the top gastroenterologists in the city. Dr. Oratz made me feel like I was with family from the first hello.

While she was very warm, she got right to the point. She explained that I would be going through several rounds of chemotherapy; it wasn't a question of whether I would have chemo or not. It was deciding which drugs I would need, in which combinations, and in what order I would be taking them. Dr. Oratz explained to me that for many years, oncologists had recommended a standard ACT regimen of chemotherapy for fighting breast cancer like mine.

The information was delivered very matter-of-factly. Dr. Oratz wasn't cold. She was professional and extremely knowledgeable about the course of action. It's wild to look back on these meetings, because now I know about all of the possible options, but at the time, I was so naive. I was gathering information and soaking it all in like a sponge. I had no idea which way was up, let alone how to choose the best course of treatment.

I was about to encounter the first of many potentially life-altering decisions I would be required to make while waging my battle against breast cancer.

Dr. Oratz laid out the same regimen that Dr. Ward had spoken of the day before, beginning with chemotherapy, which would shrink the tumor; followed by surgery; and finally, radiation. However, both doctors recommended that I flip the standard ACT treatment on its head and begin with the "T" part, a chemotherapy drug called Taxol. Dr. Oratz also suggested that we add four doses of carboplatin, the latest drug being used against TNBC, every three weeks during the twelve-week course of Taxol. Carboplatin was somewhat new to the treatment of breast cancer; though it had been used for years to treat other cancers, only in the last few years had trials been done to assess its success in the fight against breast cancer. Taxol would be followed by an eight- to twelve-week course of AC: Adriamycin and Cytoxan, which was taken every two weeks (or every three weeks). This chemotherapy would significantly reduce the size of the tumor or maybe even get rid of it.

Dr. Oratz explained that she had been at an oncology conference the week before during which they discussed the results of a new drug trial of the CALGB (Cancer and Leukemia Group B) 40603, looking at the integration of Carboplatin (paraplatin) and Avastin (bevacizumab) into the management of TNBC. This drug had been combined with Taxol for triple negative breast cancer only recently, with first reports of efficacy presented the year before, at the 2013 San Antonio Breast Cancer Conference. She outlined some pretty impressive success rates found by the doctors leading the trials of this new drug regimen.

I was learning so much about triple negative breast cancer, how the body works, how cells work in the body, and how chemotherapy works, all in such a short period of time. It was like a rapid-fire course in anatomy and biology. I experienced such an amazing intake of information. You can bet there were days when my doctors were discussing cell division and sentinel lymph nodes and I was wishing I'd paid closer attention in biology class.

The plan Dr. Oratz ultimately recommended for me was not the standard treatment used by most oncologists. My plan would have me fighting cancer for the next eight to nine months. However, Dr. Oratz assured me that the approach was, in her opinion, the latest, greatest, and most promising treatment path available for *my* needs.

That was music to my ears.

She went on to explain more about the drugs I would be taking and the side effects of the chemo. Believe me, they weren't pretty.

"Yes," she said, "there will definitely be hair loss. Complete and total hair loss."

She was saying I would be . . . bald.

Stop right there.

I suppose I knew this was a possibility. After all, I had asked Dr. Ward about hair loss when she'd told me about my tumor. But the "likelihood" was becoming something of a sure bet.

My stomach was in knots. What woman wants to lose her hair?

At the risk of sounding horribly shallow, I admit I shuddered at the thought. My hair had been my trademark for years. It was part of my brand and persona. On the day I left *Good Morning America*, the producers surprised me with a montage of my changing hairstyles over those twenty years. There were a few I'd rather have forgotten! Then again, there were a couple that women really loved. I remember being backstage at a concert many years ago where I met Faith Hill.

"I cut my hair to look just like yours!" she said.

You can bet I was flattered!

I'd heard this from women for years. And now . . . well, I simply have no words.

I get it. I know. It's hair. It can grow back. But still. I was shaken up.

The other major side effects Dr. Oratz said I could expect were nausea and a major loss of energy. I'll admit, these worried me nearly as much as the hair loss did. I'm a bit like the Energizer Bunny. The idea of not having any get-up-and-go kind of freaked me out.

Although everyone experiences the side effects of chemo treatment in his or her own way, I supposed there was always the chance that I might sail through this treatment—then again, it could just as easily bring me to my knees.

Dr. Oratz said I should begin my chemo immediately, suggesting I could take my first treatment within the coming week!

Really?

It all seemed to be happening so fast.

A little *too* fast.

I was feeling nauseated already, and I hadn't even started chemo!

The rapid pace definitely made my cancer feel very real.

At the time, Lindsay was adamant about my getting a second opinion. She wanted to know that everything we were hearing was the exact and correct course of action. Lindsay comes from a generation used to second-guessing the information received from doctors by looking everything up and researching information through various channels on the Internet. For whatever reason, this generation won't or can't accept what their doctor tells them as the gospel truth.

I, on the other hand, have always had a trusting relationship with my physicians. I've never had a solid reason not to go with whatever they tell me. I had no reason to believe anything Dr. Oratz was suggesting wasn't the best course of action. I liked her and felt comfortable at her office. I didn't want to get even more confused than I already was. But after giving it some thought and talking it over with Jeff and the girls, I realized that seeking a second opinion was the smartest thing to do.

Despite Dr. Knapp's personal relationship with Dr. Oratz, he had already set up an appointment for me with a leading oncologist at Memorial Sloan Kettering Cancer Center, one of the most celebrated cancer centers in America. Jeff, Lindsay, and I drove into New York City together and made our way to Sloan Kettering. It was a bittersweet day. Kate and Max had graduated from elementary school. Kate was the commencement speaker. She was so unbelievably poised and self-confident as she rose from her seat among the rest of the fifth-grade class and took to the stage to deliver her flawless and poignant speech. Interestingly, she never let us hear it in advance. She wanted to do it all on her own, and she knocked it out of the park. Jeff and I were so proud. We just couldn't believe that Max and Kate were out of grammar school and going into middle school. As I sat there in that school auditorium watching the ceremony, I couldn't help but think about the horrible possibility that I might die and Jeff would be left to raise our four young children alone and wouldn't have me sitting next to him at memorable life events to shoot him a smile or a knowing glance that said, "I love my life. Isn't this a wonderful moment, and aren't these kids great!"

On one hand, there we were, celebrating the kids' fantastic milestone; yet on the other, we were contemplating the unthinkable. I was so thankful that Jeff had come back from Maine to be with me for the second-opinion appointment. Having him by my side gave me a sense of calm and comfort that I don't always articulate but certainly feel when he is near.

I was intimidated upon entering the majestic yet massive New York City hospital. Though the lobby of Sloan Kettering is beautiful, I barely noticed as we dashed through it to the elevator bank. I didn't want to make eye contact with anyone. I was scared to be there.

Scared to be seen.

Scared to be found out.

Scared that I had cancer.

However, we had come to see one of the country's most renowned oncologists. There was no backing out now.

When we laid out my situation for this oncologist, he had a very different opinion as to how I should proceed. He said he would recommend starting with a lumpectomy; then he would proceed with the ACT chemotherapy, the more traditional regimen. I asked if it wouldn't be advisable to shrink the tumor first with chemo and then have a smaller surgery, since the tumor would be smaller. I shared that my cancer surgeon felt that if we first shrank the tumor, the lumpectomy would be far less invasive. He gave us lots of statistics and reasons why the tried, true, and tested way of treating my type of cancer was the better way to go.

I didn't really have the knowledge to argue the other side, so I asked him to call my breast surgeon, Dr. Ward, so we could discuss the pros and cons together. By the end of the conversation, he agreed that in my case, it made better sense to do the chemo first and then surgery when the tumor was presumably smaller. However, he still didn't agree with the new chemotherapy regimen that Dr. Oratz recommended. He felt that the more traditional ACT treatment was still the way to go, that the carboplatin was too toxic and there wasn't enough evidence to support that regimen of care. Ironically, he went on to tell us that the study Dr. Oratz was referring to was *HIS* study!

Here's why.

The use of the carboplatin regimen is still controversial. The main line right now is that it is more appropriately used in women who carry BRCA1 and BRCA2 mutations because their tumors are inherently more resistant to standard chemo. The study however, did not partition patients into groups that are BRCA positive and not. The overall result is that the response rates are higher in patients with triple negative breast cancer with the addition of carboplatin.

The end point of the research study was SURVIVAL and those data are not yet ready. As the author of the study, he could not recommend something that did not meet the PRIMARY endpoint of his own study.

While I could appreciate where he was coming from, I was more confused than ever.

Dr. Oratz was basing her course of treatment on a study that he had done, yet even the doctor who'd conducted the study couldn't necessarily go by the results.

Okay, wait a minute.

He wasn't wrong.

Hello!

Is anyone else CONFUSED?

I left that appointment with more questions than answers.

You go for a second opinion so you can hear different opinions, but hearing such different opinions from two of the top experts in their field can be very disconcerting. Each regimen would have worked. It was a matter of figuring out which approach was right for me.

How were we supposed to make this decision and move forward medically with a treatment that would offer my best chance to beat cancer? I wasn't a doctor. God, I wished my dad were around to help me make that decision.

Jeff and I discussed it and decided that, overall, I felt more comfortable with the "chemo first" approach. This treatment had such impressive success rates in the trials that it was probably the best way to go for my particular case.

I was very nervous to make the ultimate decision about my treatment, but every woman going through this journey will find herself in that situation. Again and again, she will be making decisions like this one, hoping she makes the right one. It feels like an awful game of Russian roulette that you're forced to play.

One wrong choice, and BANG!

The pressure is enormous.

Can't someone just tell me what to do?

PLEASE! Tell me what to do!

Why did I have to be in charge of these huge lifesaving decisions about something I knew so little about?

The reality is, no one wants to make life-or-death choices out of fear or under duress. Although I didn't like the way the situation made me feel, it definitely pushed me into becoming my own advocate really fast.

At one point I realized that at each juncture you come to in the breast cancer journey, you just hope that you're *choosing the path of least regrets*. I remember the first time I heard someone use that phrase. Ever since I announced that I have breast cancer, I've had a lot of touching conversations with other women battling the disease. During one chat with a woman debating whether she should take another round of chemo, I heard her say, "I'm so perplexed by everyone's opinion, but my gut tells me to go with the path of least regrets."

Good phrase, I thought. *I'll remember that one.*

A mastectomy was never mentioned by any of my health care providers as an option for me. I clearly had a tumor only in my right breast. I didn't have cancer in my left breast, and the course of treatment—a lumpectomy, chemo, and radiation—had a proven track record. While I will admit that the thought of cutting off my breasts and getting a brand-new set of beautiful boobies did cross my mind, in the end, it wasn't a viable option for me. I do understand the hysteria, how women come to think of the breasts as their enemies and they just want to get them OFF. Having a mastectomy over breast conservation HAS NEVER SHOWN an increase in SURVIVAL. A mastectomy obviously reduces the risk of local regrowth in the breast, but no reduction in death rate. It is the molecular biology of the tumor and the possibility that it could have "shot off" circulating tumor cells into the blood or lymph that can take a woman's life.

A preventive mastectomy obviously reduces the risk of getting a SECOND PRIMARY breast cancer and would have nothing to do with enhancing the SURVIVAL from the first cancer.

But in my case, there was no evidence to back up that decision. Besides that, you could cut off my breasts, but what about other cancer cells that might be somewhere else in my body?

Once we made the decision to go with the newer chemotherapy regimen, my treatment began quickly. My first chemotherapy session was scheduled for later that week.

In the meantime, my doctors suggested that I have a full-body PET scan to make sure the cancer hadn't spread to any other part of my body. Dr. Oratz said she would hate to look back someday and say, "Why didn't we check back then and find out there was something else to attack while we were taking this course of treatment?"

I must admit, it did conjure up a concern that they may find something else wrong with me, but I couldn't argue with the logic.

While I'd never had one of these full-body scans, I'd heard about PET scans and even reported on them, so I was aware that the test would show everything that was wrong inside my body—*head to toe.* I'd had several MRIs when I'd broken my shoulder falling off my horse during a jumping exercise. Also, years earlier, I had herniated my C5/C6 disc doing aerial flying maneuvers in an F-16 for *Good Morning America.* While reporting on a piece about the U.S. Air Force's elite flying brigade, the Thunderbirds, I was allowed to fly in the rear seat of an F-16 trainer jet doing aerial gymnastics. I was put through several grueling days of training before I was allowed to take to the air with the aerial gymnastic flying group to do loop da loops, fly upside down, and then the pièce de resistance: to be one of the very few civilians ever allowed to take nine G's with the Thunderbirds. (G's are the gravitational force you experience at very high speeds. Let's just say that the only people who have ever experienced nine G's and lived to talk about it are astronauts and pilots like the Thunderbirds.) The pilot I was flying with explained what we would do to create the g-force: He would bank the plane to start the gravitational force, and then I would pull back as hard as I could on the throttle to take us up to nine G's. As we began the maneuver, the pilot banked the jet and then told me to put my right hand on the throttle; he would count me down. However, when he asked if my hand was positioned on the throttle, I looked down to be sure I was grabbing the correct lever and forgot to put my head back flat against the seat, as I had done over and over in training. So when I yanked back on the throttle and the F-16 bolted into super-speed, my head was thrown violently forward, chin flat against my chest, herniating my C5/C6 disc. It hurt terribly, but A) I didn't want to look like a wimp in the eyes of the air force pilots; B) when would I ever get a chance to do this again? I couldn't simply bail out of the exercise; and C) I had no idea how severely I had just injured myself, so I made light of it.

This injury plagued me for years to come, until I finally had grueling spinal surgery to repair the damaged discs and then a fusion to secure the weakened discs to the adjoining healthy ones.

So on the day of my PET scan, while I can't say that I wasn't a little nervous, it turned out that it's really not difficult or scary. They wouldn't let Lindsay come inside the building with me because she was pregnant, and it's not safe for pregnant women to be close to radioactivity. Thankfully, Sarah had come in from Los Angeles and was able to accompany me to the appointment.

I knew I didn't have to worry about being confined to a small space, as I had been during my MRIs, because this would be an open PET scan. However, I am a bit of a needle weenie, so I was worried about having an IV in my arm or being given several shots.

When I got to the facility, I was relieved when I was merely given some medication to drink. The technician said that the liquid would light up in my body like kryptonite in a *Superman* movie. Once I drank the liquid, I needed to sit and wait for about an hour to allow it to work its way through my system.

The scanning rooms are filled with giant machines that are always freezing cold, because it's good for the machines, but not so much fun for the humans. Thankfully, the techs have realized that most people are already scared half to death, and now they're freezing their asses off! They quickly offered to cover me with a few blankets as they sent me into the tunnel and instructed me to just "follow the machine's commands."

I kept my eyes closed throughout the entire PET scan. This is a technique I use whenever I think a test, procedure, or shot might hurt or really scares me. My basic theory is "If my eyes can't see it, it's not actually happening."

Hey, whatever works, right?

Once I was lying still on the moving table and found myself inside the machine's tunnel, I heard the first command. It was a rather sultry voice that said, "Take a deep breath."

Okay—I obeyed.

Then the machine said, "Hold your breath" and once again, I submissively followed the direction I was given.

Just when I thought I couldn't hold my breath any longer, the automated stifling voice said, "Now breathe normally."

I let out an exhale of tremendous relief!

It took only about half an hour for the entire test to be complete.

When it was over, the nurse said Dr. Oratz had called and wanted me to come back to her office as soon as the PET scan was done. Lindsay met us there but was told she couldn't sit in the same room with me because I was still so radioactive from the test. In fact, we were told that when one patient walked past the UN after her test, she was so radioactive that she set off their alarms and found herself surrounded by police and Homeland Security! Apparently, you can set off the alarm at TSA,

Walmart, and lots of places after drinking the radioactive juice for a PET scan.

Great! How can that be good for anyone?

Lindsay went to a nearby coffee shop, called in to hear everything the doctor had to say, and took copious notes. For the rest of the day, wherever I had to go, Lindsay couldn't follow, but lucky Sarah—she was plenty exposed. And I am happy to report that her third arm is doing just fine.

When we got to Dr. Oratz's office, I met with Beth Taubes, an oncology nurse who would be in charge of my chemotherapy treatments, so she could give me more information on what was about to happen and what I could possibly expect from chemo.

Beth asked if I was the kind of person who felt nauseated often. I told her that I wasn't; when I was pregnant with my daughters, I often felt nauseated but never threw up. She was hopeful that I might do better than others in this category.

Then she asked if I got headaches very often. I said that I rarely did. Apparently, headaches are another common side effect of chemo, but since I wasn't especially prone to getting them, we'd have to wait and see how I'd react.

Then she wanted to know if I had trouble sleeping.

Bingo.

She had me on that one.

I absolutely had sleep issues. For as long as I could recall, sleep had been my nemesis. My body clock had been off for years. I don't know if you ever really recover from getting up at three A.M., day in and day out, for as long as I did while hosting *GMA*.

We also talked about general lifestyle during chemo, including the importance of proper nutrition. Beth and Dr. Oratz both talked to me about the concepts of eating healthy and keeping a watchful eye on my nutrition. At first I thought I understood what they meant. After all, I had been a health advocate and had written several cookbooks on healthy living! In my mind, I had been exemplary in my food choices. But really, I knew the truth. With four kids under the age of eleven, is any mother truly perfect in her food choices all the time? I wanted to nod, as if to say, "Of course! Right! Naturally!"

"No fast food. No KFC or pizza," they said. "But of course, you know all of this. You've written books on it."

Wait a minute. Were they telling me I couldn't eat cheeseburgers and pizza with my family?

Were they nuts?

I have young kids!

Next, Beth laid out more details about the course of suggested therapy and how each drug would be administered. She said I may not feel so bad the day after the infusion, but it usually hit the patient hard on the third or fourth day after the treatment. She said they would give me meds to deal with nausea and suggested other meds to deal with potential digestive issues and possible *sores in the mouth.*

What the hell?

Ugh!

I didn't see that one coming!

I was told I'd likely lose my hair fourteen to seventeen days after the first chemo treatment, maybe even within the first week, so I needed to get right over to the wig salon so I would be prepared when that happened.

Dr. Oratz handed us a sheet of recommendations for where we could buy wigs in New York City. Lindsay made me an appointment at the first place, Bitz-n-Pieces, a salon that specializes in helping women going through cancer and other hair-loss issues. They were well equipped to make me comfortable with this highly anticipated transition.

Bitz-n-Pieces is a private salon that is by-appointment-only. This ensures complete privacy for the customer, which I appreciated as I rode the elevator to my first appointment. While I watched the floors click by, I contemplated what would happen if someone inside recognized me and word got out that *Joan Lunden has cancer.* I envisioned the tabloid headline and cringed. I had been fodder so many times in the past, especially during my divorce. It had been years since my name had been linked to anything tabloid-worthy, and I'd rather enjoyed the anonymity my drama-free life had given my family and me. Still, in my gut, I knew what was possible, especially these days with cell phone cameras and instant access to the Web. Everyone has the potential to be a paparazzo these days for a price. All it takes is to be in the right place at the right time.

Sarah and I walked into the wig salon (Lindsay still wasn't allowed to be exposed to my kryptonite), and our eyes grew wide in utter shock and amazement. I looked right and then left at the vast array of hundreds of wigs, in every color and style you could imagine, on display.

I'd never seen anything like it.

It took my breath away.

If I weren't there for obvious reasons, I might have thought I'd hit the playroom jackpot. There had been many times throughout my career when I'd thought about wearing a wig—disguising myself so I could go out and be a regular gal on the town—especially when I was single and dating after my divorce.

There was one night I did don a long red wig to attend an Alanis Morisette concert at Roseland in New

York City without being recognized. I would have gotten away with it too, but the bouncer carded me at the bar! I'm not kidding, he really did, and if a guy cards you when you are forty-three years old, you seriously consider kissing him on the lips!

Barry Hendrickson, the warm and ingratiating owner of the wig salon, greeted us at the door with a kind smile and an instant assurance that I was in superb hands. Without divulging names, Barry shared that he had provided wigs to many TV and film stars—some because of chemotherapy; some because they were playing a part on Broadway or in a movie that required them to look different; and some who just wanted a wig or two for fun.

Oh, how I wished I had been in that last category!

Barry ushered us into a small private room where we were introduced to my personal wig specialist, J.T. Sarah and I loved him from the moment he said hello. I'm not sure if he was being kind or trying to make me feel comfortable, but J.T was quick to share that he had always loved the Joan Lunden hairstyle and was ridiculously excited to "Joan Lunden my wig up."

I'll admit I wasn't sure I understood exactly what that meant, but hey, I was game. And, I will concede, flattered by his enthusiasm.

With every wig J.T. handed me, he encouraged me to "Give this lady a whirl." Before I knew it, I was actually having a good time. He made what could have been an incredibly torturous emotional experience hilariously fun and enjoyable. For a brief period of time, I'd forgotten why I was there.

After trying on several "ladies" that were pretty close to my actual hairstyle, we came across one that was perfect. When he put it on my head, Sarah looked at me and said, "Mom, you look exactly like *you*."

"That's the point!" J.T. said with a wide smile, knowing he had totally nailed it.

After I tried on a dizzying number of looks, we all decided it would be best if I came back the next day with my hairstylist, Emir Pehilj. After all, Emir would be in charge of keeping my wigs looking fresh and styled so I could keep up my work schedule; he would need to be a part of this process.

Jeff wanted to drive me into New York City and go to the second appointment with me at the wig salon. The idea of this made me really uncomfortable. While I didn't mind Jeff accompanying me to any of my doctors' appointments, I wasn't sure I would feel comfortable trying on wigs in front of my husband. It wasn't like we were setting ourselves up for some *50 Shades of Grey* fantasy. In this case, it was the kind of bare I wasn't yet ready to share. The thought made the reality of being bald that much more real and a hell of a lot more terrifying.

Looking back, I think I was more scared about the thought of my husband seeing me bald than I was by any other side effect of the chemo. Despite my fumbling lame attempts to make excuses, Jeff insisted on taking me. He wanted to be by my side every step of the way. That's the kind of man he is.

Although I was reluctant, as soon as we got in the car, I felt so much better and safer with him next to me. Don't get me wrong. I was still totally uncomfortable with the whole hair-loss thing, especially the thought of my husband seeing me soon with no hair. Oh God, what a thought—but it was going to happen, and I needed to deal with it.

When we arrived at Bitz-n-Pieces the next day, the guys in the salon were so welcoming to Jeff and Emir. In the end, we all agreed on the same wig, one that looked exactly like my real hair. Jeff thought I looked completely like myself and that the wig was incredibly flattering. He said I had nothing to worry about and assured me that no one would even suspect I was wearing a wig unless I told them. J.T encouraged me to order a band that went around my head with hair attached only at the bottom so I could wear it under baseball hats or floppy sun hats when I took walks and worked out. He explained it would be much cooler on my head than a full wig during the upcoming summer months, especially in Maine, where I was still planning to be for June, July, and August.

Well, the inevitable conversation of when to shave my head finally came up. J.T., Emir, and I talked about doing it at the end of the week, when we'd be together for my final wig fitting. The more the guys talked about *how* and *when* this unavoidable event would take place, the more I realized there was a good chance that I'd chicken out.

I've done some pretty crazy things in my lifetime—adventurous challenges that would make most grown men shake in their boots, from climbing the Grand Tetons, to jumping out of planes with the Golden Knights, to training with the Navy SEALs—and yet I knew the only way I could go through with shaving my head was if I did it on my own terms. I had to just leap and not look back.

So while they talked, I pretty much just listened and nodded, knowing damn well it would never happen the way they were planning it. This was a now a covert mission that was probably best served alone.

As we wrapped things up, J.T. recommended wearing my new wig over my regular hair, "giving the little lady a whirl" at least once before I needed to wear it. "You *MUST* wear the wig," he said. "It's good to put it on and go about your day so that you will see people don't really notice a difference and won't automatically think you are wearing a wig." He felt it was important to help me have a sense of confidence and not feel "weird" when I had to begin wearing my new wig on a regular basis.

So after I left Bitz-n-Pieces that afternoon, I took the short blond wig and placed it on my head. I ran my fingers through my new chin-length hair. It was different, but not so different that people did a double take.

Chapter 5
Finding My "Why"

I have become so wonderfully, terribly aware of time, of how little of it I have left.

JANE FONDA

Actress, fitness expert, activist, diagnosed with breast cancer in 2010

Looking back, I have asked myself many times *why* the idea of going public with my cancer diagnosis was so difficult for me. I remember sitting in the doctor's office on the day when I got my results of the biopsy—the day I first heard those same words so many other women have: "You have cancer."

My mind flooded with all sorts of thoughts, or should I say questions.

Did I just hear that right?

Could they have made a mistake?

Are they absolutely sure it's cancer?

Is this the kind of cancer that they take out or the kind of cancer that kills you?

Will I need to tell people?

What will people say?

Oh, sure, like that's going to save me.

As it turned out, not so much.

For decades I've been one of those people Americans could count on to get their information from. I've been the one interviewing other people who are sick and the doctors who cure them—not the patient!

I simply couldn't understand why getting sick somehow made me feel like I was letting people down or that I had failed them or myself in some way.

Maybe it was because I have always thought of myself as a strong, take-charge, in-control-of-my-life-at-all-times kind of woman. I can't remember a time when I didn't think of myself as that person.

But this . . . Well, this threw me.

Why?

If I could understand the reason, get to the root of my feelings, maybe I could find meaning in it. There had to be purpose in my fear. If I could get to the core of that purpose, surely I could turn this obstacle into an opportunity.

Time to take inventory.

To take a look back and see where this was all coming from.

When I was in grade school, I was always a class leader. I excelled in school, so much that I was constantly tapped by my teachers to help correct the other students' papers. Yes, I'll admit, I was the one in class who always handed in perfectly bound reports for extra credit because I actually found it *fun* to research and write them. Go ahead, get it out, people like me are really annoying, right? In retrospect, perhaps that was the beginning of my desire to be a journalist.

I was also a really good talker—I mean really good! I could read a book flap and write a thirty-page report. Hey, everyone has a talent, and that's mine—or at least one of them.

I skipped my junior year of high school and started college a year early—just a few days before my seventeenth birthday. This was the mid-sixties, when young feminist women were burning their bras and hippies were lighting up their joints. Since I was underage, rather than letting me live on the campus of an American university, my mom sent me on World Campus Afloat, now called Semester at Sea, a college taught aboard a ship that traveled around the world. With only 350 students under the watchful eyes of staff and faculty, it was a much more controlled environment. My mom wanted me to "broaden my horizons," as she put it: to see the world so that I would want more from it, and so that I would someday "find a way to make a difference in it," as my dad had always professed.

And the world we did see!

Over a three-month period, we visited Spain; Portugal; Morocco; Senegal, West Africa; Cape Town and Durbin, South Africa; Mombasa; Kenya and Tanzania; Bombay, India; Kuala Lumpur, Malaysia; Bangkok, Thailand; Kobe, Kyoto, and Tokyo, Japan; and Hong Kong, China. Traveling the world like that had a profound effect on me. I believe it laid the foundation for my going into network television broadcasting. And while I'm sure my dad would have loved for me to see the world as he did, I'm not sure whether, if he had been alive, he would have approved of my going on a trip of that magnitude at such a young age. Even so, just as my father's passing did, that experience set me on a path that changed the course of my life.

After losing my father, I gained a great deal of independence yet felt a dire need to create a sense of permanent security in my life that no one could ever take away. I believe my great need for security in my life began with those feelings of insecurity I felt when my family lost our anchor and means of support. I watched my newly single mother struggle to survive during those lean years and vowed there and then to never be in that place. It motivated me to always be in charge of my own life. By the time I was in my twenties, I couldn't wait to get my career under way and to support myself. It has been that way ever since.

I've always been in control.

And now I wasn't.

In fact, this was totally out of my control.

I didn't like the way that felt at all.

It made me uneasy, uncomfortable.

Squirmy.

I didn't want everyone looking at me as a "sick cancer patient."

Heaven knows, I've never been very good with accepting help, let alone sympathy from anyone. There isn't a "poor me" bone in my body.

The mere thought of people looking at me with a sympathetic expression made my skin crawl.

"Aww, poor Joan . . ."

Ugh. No, thanks!

And yet I have always been up front about what I was grappling with in my life.

Always!

From my very first days of living in New York City and starting on the local news, my personal story has unfolded in the press and on the air. There hasn't been very much of my life hidden behind the curtain. It's been out there for public consumption and critique.

So then *why* was going public with *this* scaring me?

Dig deeper, Joan, I kept thinking.

When I first came to New York City to be a reporter at ABC *Eyewitness News* in the seventies and, later, as a reporter on *Good Morning America,* I sometimes thought I was a little like Ann Marie, the character Marlo Thomas played on the sixties hit TV show *That Girl.* She was a perky but naive young single girl in the big exciting city, stepping into the world of television news.

I was only twenty-fve when I moved to the Big Apple in 1975 and took my first job as a correspondent with WABC-TV. It was an amazing opportunity but a very intimidating move to come to the city on my own. I quickly learned my craft on the streets of Manhattan, among some of the fiercest and best competition in the business.

I must have done something right, because one year later, I was tapped by *Good Morning America* to become a part-time correspondent and then to sit in with David Hartman for cohost Nancy Dussault. After I became the permanent cohost in 1980, *GMA* viewers were with me every step of the way, just as we all followed *That Girl's* exploration of a new job in the news and a new life alone in the big city.

Thankfully, the *Good Morning America* viewers never made me feel alone on my journey. They've always responded in droves to my honest sharing of whatever was happening in my life: They sent thousands of well-wishing cards on the impending births of my three daughters, along with parenting advice and recipes for making my own healthy baby food. They knitted blankets and adorable little sweaters, and some even sent tiny handmade leather booties. The network executives took notice of this incredible outpouring of love, support, and interest and made sure to assign me as many stories on pregnancy, child rearing, healthy cooking, and living as I could do. As a result, I often interviewed leading doctors and experts on how to make the best choices to keep yourself and your family healthy and happy. The umbrella topic became my beat, so to speak. While another woman might have found it offensive to "get stuck with" or, worse, "relegated" to those "female stories," I didn't mind. As a self-proclaimed information junkie, I love perusing the latest magazines, ripping out pages with timely stories and good recipes, and I looked forward to reading every self-help book that came out. I embraced my new role on *Good Morning America*.

When you're lucky enough to work on a program like *Good Morning America*, you have unprecedented access to the latest and greatest information; it not only helped keep viewers healthy and informed, but it also allowed me to disseminate that material for myself. What a perk! The show often asked me to take part in a segment, so it wasn't unusual to see me down on the floor learning about the latest fitness craze from some hunky ripped trainer, or in the kitchen cooking with Julia Child, Wolfgang Puck, or a great new health-conscious chef putting an interesting spin on a dish that would provide you with intense flavor without the extra fat and calories.

In my early years on the job, a lot of the stories I did were focused on the best advice about raising children and how to run a home more efficiently for all those women out there who, like me, were trying to hold down a job—a demanding job—and still be good at their "other job": cooking, cleaning, and raising a family. It's a tough juggling act that many women struggle with, one that can leave you exhausted, with very little energy for anything else and no extra "you" time. That often becomes the perfect recipe for putting on a few extra pounds, which can present all sorts of health dangers that leave us frustrated and scared.

I knew all about those frustrations firsthand. You see, I had gained weight with all three of my pregnancies and never really lost it. I referred to it as my "baby weight" until, at age forty, I decided it was time to drop the weight, shape up, and get myself back into a fit state.

As at so many other times in my life, *GMA* viewers were right there to support and cheer me on. I shared everything I learned on that journey—how to create a healthy eating plan and how to find a form of fitness you enjoy so you can stick with it and make it part of your everyday lifestyle. I lost about forty pounds, and that transformation struck a chord with a lot of our viewers.

Why?

I was just like them!

We had the same responsibilities and shared the same frustrations, but I figured out how to take little steps and convert my flabby tired body into a fit and energetic one. People all around me were paying attention to what I'd done to change my routine. They wanted to know what I'd done, how I'd gotten there, and how they could do it, too.

While I didn't plan it, my career path offered me a terrific opportunity to help people make good decisions about their lives, their health, their families, and their homes. I could feel like I had a sense of real purpose in life, not just a job. And that felt really good.

I've loved my role on this earth, to seek answers of the best experts around, to connect with the audience, and to bring them the best possible information about their health and happiness. I have also shared my own life and health journeys.

And now here I was again, with an opportunity to share a journey that millions of other women deal with all over the country: breast cancer. However, being open and sharing my journey was not my first thought this time, because I didn't want anyone to see me as a failure at keeping myself fit and free of disease. I'm not totally sure why I felt that way. I think a lot of people have a tendency to equate cancer with being weak and perhaps not diligent about keeping their health in check, which of course isn't true.

For a lot of women, hearing that we have breast cancer somehow makes us feel like we are less feminine than we used to be: less sexy, less strong and vibrant, less appealing, less pretty, and I guess all about our cancer, our disease. At least that's how I felt when I first heard those words: "You have breast cancer."

How could I allow myself to feel that way? It didn't make sense!

But I did.

Still, I had always taken great pride in the interviews I'd done with health experts, and for being an avid health advocate. Not only had I headed up numerous health media campaigns, I'd spoken all over the country on how to protect your health. I founded a women's wellness getaway weekend, Camp Reveille, on the grounds of my husband's summer camp, where women are invited to scale climbing walls, play tennis, and take yoga, pilates, and Zumba while learning all about their health. Committing to this type of work has been my way of "walking my talk" over the years, especially in the years after leaving *GMA*.

So this begs the question: How could this have happened to me, as someone who generally took good care of herself?

If I went public, if I chose that path to put every nuance of my battle with cancer out there, would people be looking at me with sad eyes every time I stepped out my front door?

Would a normal trip to the grocery store become a dreaded trek?

Could I face the other moms in the carpool line at school wearing my wig?

Or worse, just a bandana around my inevitable bald head?

Would they whisper and talk about me as I pulled away, wondering what would happen to my children if something happened to me?

Would I be able to go on national television and spit those words out of my mouth—"I . . . have . . .

cancer"—without choking up or completely losing my composure?

Time had been passing at warp speed since my diagnosis. Going from one expert to another, finding out the best treatment for my kind of cancer, educating myself on my disease, doing my best to understand it all, and trying to make the right decisions every step of the way was a heavy load. Every time I walked in and out of another hospital, another doctor's office, or another lab, I was a bundle of nerves—did someone recognize me? Did they upload a picture to Twitter or TMZ?

That's the world we live in today.

And you don't have to be famous to worry about that. Facebook, Twitter—it's the fastest form of information out there for all of us.

What I had to come to terms with was if I came out publicly with the news, just went ahead and put it out there, then I wouldn't feel the need to hide everywhere I went. I wouldn't feel like I was lying every time a friend asked me a simple, everyday question, like "How are you?"

One thing was for sure: Being honest and open about my diagnosis would take away the need to sneak in and out of my weekly chemo treatments in a wide-brimmed hat and sunglasses. I mean, that would probably only draw more attention to me, right?

In the end, what I realized is that I had to choose to live in my truth, even if I didn't like it.

I had cancer.

I would be going through chemo treatments. Sooner or later, people were going to find out.

When I looked back on my life, what I realized is that I've always chosen to live an authentic life. Living anything less would be untrue to who I am. It would be unfair and unsatisfying.

So as I was trying to figure out my "why"—why I was so afraid to go public—I thought about the teachings of Nietzsche, who said that above all in life, we should always retain our authenticity and then take full responsibility for doing so. In his words, "*If we possess our why of life, we can put up with any how.*"

I didn't ask to get cancer.

Nobody does.

Okay, maybe that person who smokes twenty packs of cigarettes a week has a good idea he or she is flirting with disaster, but that wasn't me.

What I do know is that my life has always been about health advocacy, education, and information. As I faced possible death, I was questioning how I wanted to define my life. I wanted to believe that it would be about more than having been a wife, mom, journalist, and health advocate. Sure, that meant a whole heck of a lot. But there was something more, something I deeply felt connected to for years—and that's my desire to have lived a mindful life, fully engaged in self-discovery and awake to the fact that, as I have lived, I have continued to evolve and grow, as a work in progress. In the end, as long as we live, that is a choice we all have!

If I went public, put my journey out there, and shared my experiences with people, I could spread this message of hope and encouragement.

This was my *why*.

Chapter 6
Finding the Courage
to Share

*I don't tell my story to scare people but to stress
the importance of knowing your own body and
trusting your instincts.*

OLIVIA NEWTON-JOHN

*Singer and actress, diagnosed with breast
cancer in 1992*

O nce I came to terms with the fact that I needed
to step up and speak up, the time had come to
tell everyone I loved, or at least my closest friends and
family, that I had cancer.

This was a lot easier said than done.

First, I had to decide *how* I was going to do that.
Truth be told, I'd been thinking about that question

nonstop ever since hearing my diagnosis. My anxiety over facing the unavoidable conversation was growing with each passing day. I knew from the beginning there was not a chance of keeping this kind of story under wraps. If I could have, I would have put it in a tiny little box, tied a pretty string around it, and hid it deep in the ground somewhere in our backyard to protect those I loved from the pain and anguish that the news would inevitably bring. I also knew that word would eventually leak to the press. With each passing day, that likelihood grew, so I needed to tell my family and friends before they read it in the paper or heard about it on *Access Hollywood,* Twitter, or Facebook.

Jeff comes from an extremely close-knit and supportive family. They celebrate everything and cling tightly together when times get tough. As soon as Jeff and I heard the results of my biopsy and knew that we were going to be fighting a battle against breast cancer, we called his mom and dad, Janey and Donny, to let them know what was going on. Jeff's parents are young at heart, active, and a vibrant part of our family. They go on trips with us and come to every single soccer and basketball game to watch their grandkids play. They are always in the front row, cheering them on. Having in-laws whom you really like and enjoy going out to dinner with and being around is a special gift. Not only that, Janey had gone through her own battle with breast cancer years before, so we were certain she and her husband would be tremendous advocates and supporters from the start.

After sharing the news with them, we asked that they keep it quiet until we figured out the right time and place to tell other family members on our own terms.

Jeff and I had spent the day in New York City, making arrangements to start my first chemotherapy session, which would take place the next day. On our way back to Connecticut, we were going to a dinner to celebrate Janey's birthday. I knew there would be a good chance that something would come up or slip. I didn't want to detract from her celebration, but I wasn't in an especially festive mood, and I'm sure I wasn't doing a very good job of hiding it. All I could think of was how I would tell Jeff's siblings. At the end of dinner, after all of the gifts had been opened and the festivities were coming to an end, I whispered to Jeff's brother and sisters, Kip, Leslie, and Karen, that I wanted to speak to them privately and asked if they would take a walk outside with me.

We all stood on the wide front porch of the restaurant as I slowly broke my news to them. I knew that once they had the information, they would feel compelled to call Jeff and me on a daily basis to check on how I was and to see if there was anything they could do. This is just the way this family operates. If one of them is going through a crisis, they all quickly rally. Anything short of that would be terribly uncharacteristic.

I told them that the best way to help was *not* to call every day and check up on me. While I understood that this was antithetical to how they were all raised, I explained that I am not the kind of person who does well with a lot of attention at times like this. It makes me feel needy—weak. It feels overwhelming and stressful for me. I knew they were well intentioned, but I understood that if I didn't put up boundaries early on, to give myself some peace without having to answer a lot of questions over and over, I'd regret it. I told them that I was planning to go public with the news the following week. I knew I would be besieged with well wishes and questions from friends, colleagues, business partners, and the general public and that my life would likely not be any type of normal for quite some time. I hoped they understood where I was coming from. It was the best I could do under the circumstances.

This was all so new to me.

To all of us.

Aside from the family members who already knew, I had let only a few of my closest friends and advisers in on my little secret. They all agreed there was no chance of keeping it private for long and that I needed to be up front about it. They knew me well enough to understand that this was how I had always lived my life: sharing my journey with the desire to help others.

If I was going to go public, there was no question whom I wanted to talk to. It had to be Robin Roberts.

Robin had waged the same battle I now faced, and she had dealt with it publicly, with such dignity and grace. And, of course, she now sat in the seat I once held. She would understand better than anyone what I was facing, from all points of view.

I picked up the phone, called ABC, and asked to speak with Robin Roberts. Robin is just so lovely and compassionate. She immediately lent me her ear and her understanding. I had a million questions for her but tried to stay on point. I told her what I knew so far and what chemo/surgery/radiation regimen I would go through. It felt so comforting and reassuring to talk with someone who understood the journey and to compare notes, knowing that she had made it through the same diagnosis and was now cancer-free.

But as I wound down my lengthy explanation, I heard a cautionary pause in Robin's voice as she said, "It's no picnic, Joni, I'm not going to lie to you. *But you will survive this.*"

We were both quiet for a moment.

She suggested the sooner I go public, the better. "Pick a day, any day in the next week, and I will be there with you to support you." She encouraged me to share my news, to let people know how I was doing. Robin assured me that being out in front of the situation would make me feel a lot better. "The day I went public, I felt a huge weight off my shoulders," she said.

I told her that I didn't want to feel like a victim or have people feel sorry for me. I wanted to be a warrior, to help motivate other women to get their checkups every year and protect themselves from getting cancer.

We were both living examples of how women can fight and survive this kind of cancer, especially if you catch it early. I shared the story of my father and his untimely death, telling Robin that his legacy was my main motivation to help fight the war against breast cancer.

Robin said, "You tell the television audience like you just told me, and believe me, you will make an impact. You will make your daddy proud."

I knew she was right. There was no better place for me to make my announcement than among the safe and secure arms of my *GMA* family. We decided I would appear the following Tuesday, June 24, 2014, when I would let the world in on my secret: that I was one of the nearly three hundred thousand women in the United States diagnosed with breast cancer in 2014.

Chapter 7
The Day I Started Chemo

If you have a friend or family member with breast cancer, try not to look at her with "sad eyes." Treat her like you always did. Just show a little extra love.

HODA KOTB

Cohost of Today, *diagnosed with breast cancer in 2007*

I woke up on my first morning of chemo reluctant but ready to go into battle. I had never given chemotherapy a consideration, though I had worked on several media campaigns in past years when I had to study up and understand chemo infusions and the side effects that might occur. But that was research about something that could happen to someone else, a cancer patient. Not to me. It did leave me anxious, not panicked at all, just scared. I hadn't slept well, since I was nervous about the next day, but they tell you not to take any meds because they might interfere with the chemo infusion. So much for taking an Ambien.

By the time I was ready to leave the house, I had psyched myself up for whatever lay ahead. Robin had told me it wasn't going to be easy. But the things in life worth having never are.

There was just one thing I was dreading.

That damned needle.

Oh yeah, the needle that will be inserted in my arm so the doctors can start the infusion.

I know I've already mentioned this, but I DESPISE needles!

No, I mean I really, *really* HATE them!

You see?

That's how much I can't stand them!

I know it's a learned fear—one I've lived with my whole life—and one I've tried not to pass on to my children, but it's one I have nonetheless.

It used to be a lot worse. After many hormone shots during several rounds of fertility treatments, and then two years of allergy shots, I actually got a little better with needles.

The only good news I got when I was diagnosed with cancer was when the oncologist said I had to stop getting my allergy shots! My fear of needles was the reason why I didn't get allergy-tested for the longest time. I knew I would be told I should be taking allergy shots, and I didn't want to. When I did get tested, of course I was told that if I really wanted to be cured from all the allergies that had plagued me for years, I needed to take two shots in each arm once a month. While they have helped tremendously, fortunately, I thought to call and ask my oncologist the day before chemo if I should be going in for my scheduled allergy shot, and the answer was a resounding no.

Whew! No more allergy shots for a while! I thought.

And even though the thought of not having to deal with my allergies for a while might have been appealing, that quickly faded when I discovered the allergy shots aren't nearly as bad as starting an IV in the veins of your arms every week for chemo.

That *really* flipped me out!

Jeff and Lindsay were with me for my first chemotherapy treatment. Every day I counted my blessings to have such love and support all around me in the form of my family and friends. There is no way I could have made it through without everyone rallying around me the way they did.

I took a deep breath when the nurse said it was time to get started. Sure enough, it took a few stabs in the arm before the oncology nurse found a viable vein to start my first chemo infusion. Naturally, this would happen to the ultimate needle weenie on her first day of chemo.

Once the nurse finally got the needle in my vein, the rest of the treatment process was relatively easy. I pretty much sat back in a big easy chair and let the chemicals drip into my body for the next several hours. There were pre-meds dripped in first, followed by anti-nausea drugs, steroids (which would hype me up for days), and then Benadryl, which made me really drowsy for the rest of the treatment. Benadryl (also called diphenhydramine) is given as part of the pre-medications to prevent possible hypersensitivity infusion reactions to Taxol—and it is also sometimes used in anti-nausea regimens. Taxol, or more specifically the carrier that the drug is in, called Cremaphor, can cause itching, hives, or even wheezing. The dexamethasone, ranitidine, and Benadryl given on Taxol days are to prevent this reaction.

I'd taken along my Kindle, thinking I would get through lots of reading, but the next thing I knew, thanks to the Benadryl, it was lights out—I quickly fell asleep!

Much to my surprise, the overall first-day experience wasn't that bad. Now I would have to go home and see how the Taxol affected me.

For my first treatment, my doctors had made a decision to start with only the Taxol and not the more potent medication, carboplatin.

Once we got back home and I got settled in, Jeff had to reluctantly pack his car and hit the road for the six-hour drive back to Maine. Although I was now deeply immersed in my treatment, he had to resume all of the tasks involved in opening up his summer camp. I knew he felt so conflicted, which was the last thing I wanted him to be dealing with. I could see it in his face, in his eyes. Leaving for Maine was such a tough dilemma for him. I know what it's like to have to be somewhere else when you want to be home with your family. I'd spent years having to go on the road for *GMA* and be away from my kids when they had bouts of the flu or chicken pox. My heart sank every time I couldn't be there for them. It never got any easier with time, either. I worked very hard never to miss important moments—school plays, birthdays, dance recitals, horse shows, and never, ever a parent-teacher conference. And I did my best to be home as often as I could be, though my job called me away at times. Now here I was, dealing with this unexpected disease at the most inconvenient time. And though I knew Jeff didn't want to, he had to leave. The only silver lining was that I would soon be heading up to Maine with our four youngest kids, whose last day of school was a day away, so we wouldn't be far behind. That surely made everything a little easier.

There was one thing weighing very heavily on me. Although I had known about my diagnosis for ten days, by the time we were leaving for Maine, I hadn't told the younger kids about my cancer yet. I don't think they suspected anything was wrong, because Jeff and I kept everything startlingly normal. For a moment, we even contemplated whether we could get away with sending the kids to camp without telling them anything at all. Of course, that was just a thought. We knew we had to be honest with them, because if they found out the truth while they were at camp, they'd never trust us again. We'd always been forthcoming with our children, so despite the nature of the news and my maternal inclination to protect my children from anything that would hurt them, I had to be honest about my diagnosis, too.

Things were happening so fast, but we wanted them to finish their school year with all of the happiness and joy they deserved, especially Kate and Max, who were celebrating their birthday in the midst of this chaos. I didn't want anything to detract from their celebration. But I also knew the word of my diagnosis would be slowly leaking, and with my impending public announcement, the time had come to sit them down and tell them the truth. I didn't have it in me to do it on my own. I was scared to tell them. As a parent, you want to protect your kids from everything bad in life, but especially from something like worrying about Mommy having cancer. I definitely needed Jeff by my side for this one. It made a lot more sense for us to talk to them as a family, and therefore, it would wait until we were all in Maine for the weekend.

Once we arrived in Maine, I couldn't stop moving. Seriously!

I was running around like the Tasmanian devil from the steroids they gave me during my chemo treatment, nervously trying to get all of us settled in for the next three months. I was a little surprised at how good I felt—like I could take on the world. But I'd had only one treatment. I had no idea what was coming. As soon as those steroids left my system, I had been warned, I could be left feeling crummy.

Max, Kate, Kim, and Jack would be leaving us in a few days to spend the rest of the summer as campers, which gave us only a few days together before we all went our separate ways. They wouldn't be back in the house until the end of August. We would have the conversation over the weekend, so we would have a few days with them before they moved into their camp bunks. That way, if they had any questions, we would be able to answer them.

To make sure I broke the news to the kids in an appropriate way, I had asked Dr. Oratz and her staff for advice on how to handle telling young children. They said that being honest was the only way to go. Child psychologists warned against trying to keep cancer a secret, because family secrets are a lot harder to keep from children than one might think—children sense that something is going on, and unfortunately, their "imagined" problem is usually worse than the truth. The experts also pointed out that not being honest with children can create trust issues later on.

Jeff and I called the four kids into the living room and sat them on the sofa for a family meeting. Jeff did the talking, as he is an expert at talking to young children and so much better at communicating these types of things than I am. He had given a lot of thought to how to discuss it in a very age-appropriate way. I was afraid I might break down, and that was the last thing I wanted to do in front of our children.

Jeff told them about my cancer diagnosis and said that the treatment for it had already begun. He explained it would be going on for several months. He reassured the kids that I had a very good prognosis because we'd found it early and could treat it quickly.

Telling your children you have cancer is a parent's worst nightmare.

Oh my God, how I wished this were a dream I could just wake up from.

But it wasn't.

Cancer was my new reality.

To be certain, cancer is a super-scary word for kids because it can mean so many things. We didn't want to frighten them; we needed them to understand my cancer was treatable. They all listened intently, with wide eyes, as Jeff explained that my treatments were going to be very strong, and because of that, there would be some side effects.

Kate asked, "Will Mommy lose her hair?" Kate is eleven going on sixteen, an old soul and mature beyond her years. She is very smart, and I knew she would understand the immensity of this family meeting right away. It would be difficult to ever try to pull the wool over the eyes of Kate.

"Mommy will lose her hair, and she may not be feeling so well," Jeff responded in a soft and loving tone.

Then Max asked, "Is Mommy going to be okay?" Max is the most loving little boy. He is very sensitive and affectionate. I knew he would likely be the most worried. I didn't want to let him see how nervous and emotionally fragile I was, especially early on.

Jeff reassured Max, "Mommy will be fine." He told all of the kids that the process would take time, and we would give them constant "Mommy updates" so they always knew what was going on.

I had a huge tight knot in my stomach and a giant lump in my throat, and it was really hard not to get teary as I listened to Jeff talk with the kids. It's a delicate balance to know how much young children can absorb when you tell them something like this.

By not crying, were they being brave for my sake?

Jack and Kim, at only nine, sat silently and didn't say anything. I wasn't surprised, though. Kim is usually quite shy and quiet, although she is blossoming into a beautiful young lady, and she'll get up on a table and belt out a song at home with the family or lip-synch and dance with her sister to the latest pop tunes when no one else is looking. Jack is a very sweet, compassionate boy who is always concerned about every other kid in his class. He is quick to run and get me anything I need, even when I'm sick, so I knew he would want to be attentive, though he was sitting quietly at that moment, taking it all in.

I suppose kids will be kids, because as soon as our family meeting ended, we were all out the door to dinner. We thought a lighthearted dinner would balance the heavy-hearted conversation we'd just had. And it did.

A few days later, when camp opened for the summer, it was business as usual. The kids were off to join the other children and find their bunks, and Jeff went into high gear heading up what is essentially a summer resort for children. It's a huge job, overseeing several hundred children ages seven to fifteen, as well as several hundred counselors and grounds staff and kitchen staff, and safeguarding the lives of everyone. I'm always in awe, as he is such an amazing leader and role model: always calm, always fair, and always understanding of each and every one of the campers and counselors. I don't know how he does it with such attention to detail, but he is truly masterful.

One thing I knew for sure: I didn't want my cancer to distract or upset him. The last thing I wanted was to be a burden. In truth, I didn't want to be a burden to anybody.

Ugh!

The thought caused me so much stress.

I'm a pleaser.

I'm a fixer.

I'm a doer.

I'm a do-gooder.

I'm a loyal, hardworking soldier.

I'm a mom.

I'm a wife.

I'm a caretaker.

BUT I AM NOT A BURDEN TO ANYONE.

This was perhaps one of the scariest parts of knowing I would be battling cancer and everyone else knowing that I was fighting this battle.

I hated the thought of having to ask people to drive me to my chemo sessions and doctors' appointments.

I hated the thought of my friends and business partners feeling like they always had to reach out to see if I was okay.

And I hated the thought of Jeff being deluged with questions about how I was doing in the middle of his busiest time of year.

I wanted to scream from the highest mountain, "Hey, everybody, I will be okay, and in the meantime, I'm a little busy making sure that I'm going to be okay."

Okay?

I know that might seem a little over the top, but it's how I felt.

And I know from so many of the letters and emails I got from other people that I wasn't alone in this feeling.

When you've lived as independently as I have for your entire life and you suddenly lose that freedom of

choice, it challenges everything you know and understand about yourself. I wasn't wallowing in self-pity
or feeling bad about my disease. Quite the opposite: I
wanted to ignore it. To stick my head in the sand and
pretend it wasn't there. But no one would let me do
that. I was constantly reminded of its presence.

Not by looking in the mirror and seeing my bald
head.

No. That would have been a harsh enough reminder.

Not by my lack of usual boundless energy.

Not by the chronic heartburn I hadn't had since my
last pregnancy, nearly thirty years ago.

But by the daily reminders that would come from
people being kind, being lovely, being thoughtful,
being caring, and reaching out, simply wanting to
know, "How are you?"

I just wanted to be normal.

I just wanted to live as I always had.

But my normal had changed.

I was the one who had to get used to that.

This was my adjustment to make.

Not theirs.

Still, I wanted to put out a memo that simply stated:
I'm fine!

I'm going to be okay.

And I will get through this.

Chapter 8
Good Morning America, I Have Cancer

I'm stronger than I thought I was. My favorite phrase has been "This too shall pass." I now understand it really well.

ROBIN ROBERTS

Cohost of Good Morning America, *diagnosed with breast cancer in 2007*

My appearance on *GMA* was fast approaching. The reality of going public was creating an odd and unexpected pressure I hadn't anticipated. It made everything feel like it was moving in even higher gear. The day before the show, Lindsay helped me compose an email that would go out to my many friends and family members so they wouldn't feel sideswiped by the news of my illness. I wanted them to know what was happening before the rest of the world got wind of my diagnosis. There was so much going on all at the same time. Plus, chemo was already taking a toll on me. I was in such an emotional state. I couldn't get my thoughts straight, and Lindsay knew my voice better than anyone.

After covering many stories about breast cancer over the years, I was now going to be the cancer story. Since the moment I was hired as the cohost of *Good Morning America*, I have lived my life out in the open, sharing my many joys and, yes, my disappointments with the world. I have shared my pregnancies, my relationships, my weight gain and weight loss, and throughout the years, my ever evolving career.

Yes, I have shared my entire journey.

So it definitely didn't feel right keeping this part of my journey a secret.

There is another important reason why I felt it was absolutely necessary to share the journey with everyone. I thought about what my dad would say if he knew that I was going through cancer right now and wasn't using my voice to inform and educate. I knew in my heart that he would expect me to use my journey—come what may—to inspire others to get screened for all types of cancers. Having operated on many cancer patients, he knew that early detection was crucial. I remember listening to him from the sidelines when my mother, brother, and I would accompany him to cancer conventions where he spoke about the complex nature of cancer surgery, extracting the deadly tumor while preserving the integrity of the patient's body. I was very young at the time, and my mom probably didn't even think I was listening, but I was. I was always in awe that he saved people's lives. I remember him saying many times, "If only we had caught this sooner."

I considered myself extremely fortunate that my cancer was found in the early stages and my prognosis was so promising. I wanted everyone to know that breast cancer is not something to be ashamed of or something that we should feel is taboo to discuss. If I stood tall and spoke about my cancer, it would help others speak about theirs.

Yes, I must share this story.

Here's an excerpt from my letter:

I have already begun my chemotherapy and I have been blessed to have my husband, Jeff, and my three older daughters with me every step of the way. I am so thankful to have the support, wisdom, and guidance from all of my doctors and the loving support of my family and my friends. I know I have a challenge ahead of me in this journey; however, I have chosen to see it as an opportunity to fulfill my father's legacy and try to inspire others to protect their health.

I wanted to let you know that tomorrow I will be back on *GMA*, to make the announcement.

When Lindsay and I finished putting the final touches on this very personal email that would go out early the next morning, we drove to New York City to spend the night at her apartment and eliminate the one-hour drive from Greenwich in the morning. Lindsay's husband was away on business, so it would be fun having a "city sleepover" with my grown daughter.

It had been several years since I'd lived in Manhattan. I'd forgotten how loud the sounds of the city can be at night. A fire engine racing down Sixth Avenue, just outside Lindsay's eighth-floor window, at three A.M. was no longer something I could sleep through. Though I did get a little bit of shut-eye that night, I can't say it was more than a few winks between tossing and turning.

At five A.M. Emir arrived to do my hair and makeup. Perhaps for the last time . . . at least for a while. Boy, oh boy, that was a sobering thought. I sat in a chair for two hours and let him do his best to make me look like I'd actually gotten some sleep. Losing those dark circles under my eyes gets harder and harder as I get older.

Does anyone else feel like this?

Maybe it's just the lighting in Lindsay's apartment.

Yeah, that's it.

Can I blame those on the chemo yet? I wondered.

At seven-fifteen a black Town Car came to pick us up. Rolling through the city reminded me of my long stint at the early-morning show, except I rarely saw the city streets before dawn.

I knew I would see lots of old buddies who still worked there, including cameramen, stage managers, bookers, and others who'd been around during the days I'd sat in the host seat, but this time I was there under rather unusual circumstances. I purposely hadn't done hair and makeup there, because I'd known what would happen. All of my cohorts would stop by to say hi. On that particular morning, I was worried about any potential emotional exchanges. I wanted to be empowered and strong when I went on the air.

My biggest fear was that I would get mushy and cry. *NO! I will not do that,* I kept saying over and over in my head. I needed to be strong for all of the women out there dealing with this disease.

I needed to be a role model.

I needed to be encouraging and inspiring for every woman who hadn't yet been screened, or who had felt a lump and hadn't gone to the doctor because she was afraid of the answer she might get.

When the car pulled up to the studio, Patty Neger, a longtime friend who had been a booker at the show since before I started there, met me at the stage door. She knew about my cancer. I had shared the news with her the day I spoke with Robin, since she was the point person who would arrange my appearance without letting anyone know why I was coming on the program. Patty was the person who always booked the most prominent and brilliant doctors on *GMA* to discuss the latest news about cancer and other diseases. She knew all too well how serious this was.

There are always a lot of fans waiting out in front of the Times Square studio entrance to see who gets out of the cars. As I walked into the studio, I heard a murmuring among the crowd: "That's Joan Lunden."

Little did they know *why* I was there that early, humid summer morning.

Although I'd had only one treatment, the chemo was already having an effect on my hair—it looked a little like I had walked through a sauna on my way to the studio. I felt like I didn't look my best, but I had to get that out of my head. I was totally unnerved as I walked into the studio; it was like an out-of-body experience as staffers who knew me smiled and greeted me, unaware why I was there. It was such a relief when I caught sight of Jill Seigerman, one of my closest friends, who of course was one of the few who already knew why we were there and who had worked side by side with me for eight years at *GMA*. Jill had lived through a lot with me, including my final days at *GMA*, and she wanted to be right next to me that morning, knowing I might need some moral support. A few minutes later, we were joined by my daughter Jamie and my publicist, Stan Rosenfield. Stan has also been at my side for years, protecting me from the press (or at least trying to) through my divorce, my reentry into the dating world, remarrying, and having twins with the help of a surrogate. He'd been through the wringer with me, and we were both still standing. I was awfully glad to see him, though I wished it had been for a happier occasion.

Emir applied his last-minute touch-ups, the soundman came in to mic me, and then I made my way down to the studio. Amy Robach, who read the news for *GMA*, immediately came over and gave me a hug; she had obviously been filled in, since she was going through the same thing. Her long hug was clearly from the heart and immediately put me at ease.

Then I caught a glimpse of Robin, who was waiting for me on the set. She is an incredibly beautiful woman, inside and out. She exudes compassion and heartfelt warmth.

Within seconds of me sitting down, we were on the air.

There we were, two pros, doing what we do.

Only this time, the tables were turned.

The interviewer was now the interviewee.

Three . . . two . . . one . . .

Robin introduced me and asked that first question: "It's said the people who stand by you are your family. And you all know and love Joan Lunden, long-time cohost of *Good Morning America* all those years with Charlie and Spencer. Well, she's been a part of our extended family and always will be. She's chosen to come here this morning to share something personal with all of us. It's great to see you, Joni . . . What is it you want to share with me?"

I think my voice broke a little when I began telling Robin and the viewers at home my truth. "I've covered many stories about cancer, but somehow I just never thought I'd hear those words that every woman fears and never wants to hear: 'You've got breast cancer . . .'"

I proceeded to tell her and the rest of America my story. Robin and I also spoke about the importance of self-breast exams, of getting regular yearly screenings, and of asking your doctor if you should be getting the ultrasound and other information.

She then asked about my prognosis.

I said, "Thankfully, I caught it early, it was only Stage Two, so the doctors feel that with chemotherapy, surgery, and radiation, I should beat it." I explained that I had already begun my chemotherapy and how, ironically, she and I not only had the same kind of cancer, we had the same oncologist as well.

You could hear a pin drop in that studio.

Many of my former *GMA* colleagues had been in other production rooms when the interview began and had come out to the studio to give me a hug and say how sorry they were for me.

Oh, shit.

There it was.

It had started, that dreaded "poor Joan" stuff I didn't want from anyone. I could see it on everyone's face, and there was nothing I could do about it. The proverbial cat was definitely out of the bag.

Now *everyone* knew.

Chapter 9

#TEAMJOAN:
An Unexpected Outpouring
of Support

Cancer didn't bring me to my knees. It brought me to my feet.

MICHAEL DOUGLAS

Actor, diagnosed with throat cancer in 2010

After I made my announcement on *Good Morning America*, you might have thought that I would take a big breath and exhale. After all, the truth shall set you free, right?

Instead, I felt like I was holding my breath. What would be the public's reaction?

Would people think less of me?

I didn't want that.

Would they feel sorry for me, the sick cancer patient?

I definitely didn't want that!

Was I afraid of how they would react?

Or was I afraid of how they may *not* react?

Would they be supportive?

Maybe no one would care at all.

It had been a while since I'd connected with the *GMA* audience.

When I'd last hosted the morning show, the Internet had been in its infancy. I had no idea the power or influence the Web could have with an announcement like the one I had just made.

I hadn't given a lot of thought to just how much our world had changed with the connectivity that came with social media until I got home from my epic day. That was the first time I checked social media to see what kind of response I'd gotten to going public with my news.

Thousands of people were connecting with me to send their well wishes. Some were people I'd known through the years and had lost touch with, but mostly, they were people who watched me on *GMA* over the years. Not only were they connecting, but they were sharing their stories with me. Some of those stories were that they, too, had been battling cancer.

When I was on *Good Morning America,* I was aware of the millions of Americans out there, but there was no real connection; they could call the network and write us letters, but they couldn't talk back to us.

Well, they sure could now!

This life event—getting cancer and going public—brought me together with so many of those who used to be "on the other side of the camera," and in a very kindred way, I loved it.

I loved hearing from everyone, getting to hear each and every story, like that of Cindy G.

I have known you (on TV) since the '70s. Watched you every morning before college, then when feeding babies, then before work. It seems we are about the same age. Every time you went through a life-changing event, so did I! We were pregnant the same times, divorced the same times, looked for a place for Mom with you, and so my point is that I have a real connection with you.

I live in Georgia (the Bible belt) and have many prayer warriors who join me in praying for your strength and healing. Hang in there, kiddo!

Truly a Fan,

Cindy G.

What struck me most about Cindy's message was how she felt like she knew me well, even though we had never met. I was hearing this sentiment in message after message, unlocking a door to reveal a huge party of friends through the power of the Internet. It was a new connection I could not only see but also hear. It had been eighteen years since I'd said goodbye to these morning friends, and I had missed them. I was so happy they had returned at this moment in my life.

Just wanted to tell you that I am thinking about you! I am celebrating my one-year "cancerversary" today. I, too, was triple negative . . . stage 1b, one node involved. Had a double mastectomy, chemo, and radiation. You may never read this, but if you do, and I can help answer any questions, please ask away! Be strong and positive! You can do this . . . it's a blip on the radar!!! Big hugs to you!!!
Cathy W.

Joan—
I saw your interview with Robin Roberts this morning and just wanted to send a quick note to wish you well through your breast cancer journey and thank you for taking your story public, even though I'm sure it was a difficult decision at first. I am a sixteen-year breast cancer survivor (trust me, you like people to tell you those things at the

stage of the journey you are now!)—my children were thirteen and nine when we had to share the scary news. No doubt breast cancer affects every member of your family . . . and it is obvious you are blessed with an amazing support system. I have no doubt you will win this battle. And your story will impact so many in a positive way. Just remember to breathe and pamper yourself regularly! My good thoughts and prayers will be with you (along with thousands of caring friends and fans of yours around the world). My wish for you is a speedy, successful passage through the surgeries and treatments—and that you "get to the other side" of this nasty thing called cancer very quickly. All the best to you and your family,
    Judy S.

Here is a message from Kathy Jackoway, who was my assistant at *Good Morning America* in the mid-eighties before moving to California to marry and have a family. She reached out to me on Facebook. I hadn't heard from her in years.

Hi Joan,
I am so sorry to hear the news. You are so strong and positive and I know you will be okay. I had a

double mastectomy in January. I tested positive for BRCA2 after my sister was diagnosed with Stage IV breast cancer. I had planned prophylactic surgery, but as it turned out, my pre-surgical testing showed that I already had breast cancer . . . Anyway, I just wanted to let you know that I'm thinking of you and sending you lots of positive thoughts and prayers.

xoxo

Kathy

After reading the messages flooding in on social media and emails to my website, I started to realize that there was quite a sisterhood among women battling breast cancer and breast cancer survivors out there. If I wanted to, I could turn to a lot of groups for advice and support. That was comforting to know so early in the game, especially for someone like me, who loves gathering information.

Dear Joan,

I know we will probably never meet. I want you to know how much it meant to me when you accepted me as a friend on Facebook. My son Kevin had just been released from intensive care with a diagnosis of Addison's disease. He had been diagnosed at age twenty-one months with diabetes and I thought this new disease was beyond unfair. Then somehow I saw you on Facebook and you accepted me as a friend and for reasons I can't explain, my outlook on life improved. I went from beyond devastated to cautiously optimistic. You reached out to someone not knowing how much it could possibly mean. Now I would like to do the same for you. I am going to keep you in my daily prayers and ask God to heal you completely. I have a friend who is a priest and my dear departed aunt was a nun for seventy years. We are going to storm heaven with prayers. So take care, my dear friend, and fight the good fight and leave the praying to others.

Kristina W.

As a show of support, *GMA* posted #TEAMJOAN on its Facebook and Twitter pages.

As the day progressed, Lindsay checked my Facebook and Twitter pages and found an avalanche of well wishes, along with all sorts of tips and advice on how to deal with breast cancer and chemo. In fact, Lindsay told me I was *trending* on Twitter for most of the day. I can't say that I totally understood what that meant at the time, but she told me it was a good thing. Here are just some of the amazing tweets I received from friends and colleagues that day:

Katie Couric_@katiecouric Dear@JoanLunden, thinking about you and sending positive thoughts and love your way. xoxo

---

Leeza Gibbons @LeezaGibbons @JoanLunden@gma @Robin Roberts Family of Strength and Support! Joan we love you and bless you on this journey. You ARE a #Warrior & #Winner

---

Susan G. Komen @SusanGKomen Lots of love to @JoanLunden former GMA co-host who revealed this am that she is battling BC. Truly inspirational.

Stand Up To Cancer @SU2C We applaud @JoanLunden for her courage, her openness, and the inspiration she provides us all. Joan, we stand with you in this battle.

LIVESTRONG @livestrong @GMA@JoanLunden @RobinRoberts sending our best wishes and strength to you Joan. If we can help in any way, don't hesitate to reach out.

DeborahNorville @DeborahNorville Just sent a long note to @JoanLunden wishing her a speedy recovery as she fights #breastcancer. Was just w/ her.

Tony Perkins @TonyPerkinsFOX5 Thoughts and prayers to @JoanLunden @GMA on her battle with breast cancer. With your amazing inner strength, I know you will beat this.

Cynthia McFadden @CynthiaMcFadden @JoanLunden sending all love and support!

Nancy O'Dell @NancyODell Journalist Joan Lunden announced today that she has been diagnosed with breast cancer. Lunden hosted Good Morning America from 1980–1997.

Kathie Lee Gifford @KathieLGifford_ @JoanLunden sending you my love and prayers as you deal with this latest challenge. God bless you and the family.

Barbara Walters Joan dear, you are in my thoughts. I wish you well. You are a wonderful woman. I send you a hug of love. Barbara

I was floored by the overwhelming response and warm reaction of the *GMA* viewers. I had no idea there would be that kind of outreach and outpouring of love, support, and kindness in the hours, days, weeks, and months that followed the broadcast. None whatsoever.

To be very candid, I felt amazing relief, too. I hadn't recognized that although I had been off the air for years, there were so many people out there who cared enough to send me words of strength and love. It was an indescribable, wonderful feeling that took my breath away.

I knew in my heart of hearts that if this was the response from the announcement on *Good Morning America*, there was a good chance that I could help raise awareness about early detection; start important conversations; and help make a difference. And if that was the case, going public had definitely been worth it. I'd had no idea how many lives my story would touch. There was a need for a leader in this discussion. Hopefully, I could become a source of strength and a role model for other women to get checked, to get treated, and to survive breast cancer with grace and dignity.

What I immediately understood was that cancer did not need to define me, but how I lived with cancer and fought my cancer certainly could.

This was a real revelation and, as Oprah would say, an "aha" moment for me. It was also the exact moment when my whole attitude began to change and, without a doubt, strengthen. I had a rare opportunity to use my cancer as a platform to help others. In the process, I might help save lives.

As a result of this understanding, I began writing my "Breast Cancer Journey" blog. My hope was that in sharing my journey, I could help other women approach and manage theirs. I also thought writing might turn out to be a therapeutic outlet for me.

After I shared my first blog post, I had no idea what kind of reaction I would get.

Would anyone read it?

Did anyone care?

Would I get advice or questions from other women?

Would I be able to offer advice or lend a hand to help other women?

The responses started coming in, and once again, I was stunned. The reaction was beyond my wildest expectations. It was exciting to be corresponding with others sharing the same experiences, or those who had been there, done that. While I didn't have all of the answers, my blog created an instant forum for people to talk and share. And sometimes that's enough to get the conversation started.

Here is just one of the comments I received in the days after I shared my initial blog post:

Jerri Smith has left a new comment on your post "I Have Breast Cancer"

Joan, in May of 2006, I was diagnosed with Stage III breast cancer—not one but two lumps in my left breast (I've always been an over-achiever). HA! After two surgeries, six months of chemo, thirty-three radiation treatments, and an entire year of Herceptin infusions, I'm still here—a loud and proud survivor. You can beat this—my mantra was "It's only cancer."

Don't give it any power, and you will win. Cancer isn't for sissies, and you'll have good days and bad days in the upcoming months, that's for sure. But in the end, it's your attitude that will get you through. Be tough and find your inner strength. Also, maintain your sense of humor. BTW, in 2008, I was diagnosed with endometrial cancer. Another surgery, and again I'm cancer-free. Hang tough—you can do this!

Getting those types of letters almost made me feel good about the whole thing.

Almost.

I also got a surprise message that Jim Kelly, the retired Buffalo Bills quarterback, was trying to reach me. When I told my husband that Jim Kelly had called, I think he was a little jealous! He said, "Do you mean *the* Jim Kelly, former Buffalo Bills quarterback and Hall of Famer and one of the greatest football players ever to play the game? He's one of the greatest athletes who ever lived!"

"Yeah, that's the one," I said.

Of course, I had no idea.

Before calling him back, I did a little research so I wouldn't be totally out of the know when we spoke. I found an article on ESPN.com that read: *Jim Kelly was declared cancer-free Thursday after biopsies came back negative.*

Kelly had completed chemotherapy and radiation treatments for cancer in his jaw. There was a picture of him as he left the hospital, with friends and family and teammates all lined up to give him a high five!

The article went on to say:

That scene is as indelible in Kelly's mind as any touchdown pass he ever threw, any heroic last-minute game-winning drive he engineered. It is a moment that will stay with him for the rest of what he hopes is a lengthy life.

Kelly said he just wasn't expecting it, and that he felt overwhelming joy and a spiritual uplift from that act of kindness. "It was a surprise. It was my last radiation treatment, so I figured go out and get in my car and go home, and I walk out there and they set me up. That was a point that made me feel that people really do care. I mean, I knew that they did, but to have some of those people come out, and not know they would be there, it was nice."

With that, I gave Jim a call, still uncertain why he was trying to reach me. Maybe he had started a charity that he wanted me to work on? As it turned out, Jim didn't want anything from me. He shared that he'd really struggled with going public about his cancer story at first because he was such a private person. He didn't want to feel like he'd let people down.

He said he felt like he had always been a strong athlete for people to look up to and admire, and he'd worried about what they would think. His wife had tried to tell him that he would be making a difference and could inspire people by sharing his story.

Jim said that he had been overwhelmed by the joy of all those who reached out to him and sent their well wishes. It seemed like everywhere he went, people stopped to ask how he was doing. Ever since I'd gone public, people not only wanted to ask how he was doing, they immediately started talking to him about how "Joan Lunden's going public is helping so many other women."

Hearing that over and over again helped Jim understand that there are those of us who have a special opportunity to make others feel better during a difficult journey. He said that after months of struggling, he finally "got it."

By sharing his story, he had started to see the significance of his own positive impact.

The reason for his call?

He simply wanted to thank me for helping him understand that going public meant something to others.

"Well, Jim, in helping others, we often help ourselves," I said.

I hung up the phone and smiled.

During our call, Jim had confirmed something important for me, too. Not only could sharing my story *help* others, it could *inspire* them, too.

Chapter 10

Going Rogue—Make Me G.I. Joan

I've always thought of myself as being a warrior. When you actually have a battle, it's better than when you don't know who to fight.

CARLY SIMON

Singer/songwriter, diagnosed with breast cancer in 1997

My day started out like any other. I had my usual checklist of errands and some personal appointments to get done before heading back up to join Jeff in Maine for the rest of the summer. This was going to be kind of a "beauty day" for me.

I'd say I earned one!

But I also awoke feeling as though I needed to kick myself into *warrior* overdrive. The emails, tweets, and Facebook posts from everyone had motivated me to fight the good fight against my disease harder, stronger, and fiercer than ever.

What better way to shift into warrior mode than meeting my best friend, Elise Silvestri, for a manicure and pedicure! I first met Elise when she was my intern in 1979. I was a cub reporter at *GMA*, and she was still in college. When I became the cohost of *GMA* in 1980, Elise became my assistant. The two of us were stepping into the world of big-time national television at the same time, not knowing anything about how to navigate it. I guess you could say we grew up together in many ways.

It was Elise's first job out of school, and I was starting my job as cohost with a brand-new baby, my oldest daughter, Jamie, eight weeks old. Elise and I had some amazing years together, especially during my early days at *Good Morning America*, and have remained best friends to this day.

Elise was concerned to hear about my cancer, and we had made plans to meet and catch up on everything that had happened since my diagnosis and going public. While we were being treated to hot-rock leg massage as part of our "warrior mode" pedicure, I leaned over and whispered to Elise that I was wearing one of my new wigs. "What do you think?" I asked with a smile.

She was stunned that it was so natural-looking!

Hey, if my best friend hadn't noticed, no one would!

Okay, so J.T. was right.

I also told her that I was likely to be losing my own hair sometime around the Fourth of July weekend, when I was expecting a houseful of guests for the holiday. It scared me to think that chunks of my hair might come loose or fall out while I was serving lunch, or worse—right in the middle of a conversation. That's when I told her that I was seriously considering shaving my head . . . later that day.

I mentioned that Robin Roberts had recommended getting in front of the hair loss by shaving it off before it fell out. "That process," Robin said, "can be depressing and embarrassing."

I felt a strong need to take control of it and do it on my terms. I had a spray tan scheduled for that afternoon. I said I was thinking of making *a bold bald choice.*

I mean, come on.

If you're going to be bald, at least look healthy and tan from head to toe!

Am I right?

I had just taken care of my toes at the nail salon, so what about my soon-to-be-bald head?

I figured the only thing worse than a bald head was a pale pink bald head.

I was doing my best to find the humor in this otherwise awkward situation. I wasn't sure whether Elise was worried or amused.

I asked her what she thought about the idea of me just walking into the beauty salon I normally went to for my spray tan and asking one of the hairdressers there to shave my head. Although I'd never had my hair done there, did it really matter? It wasn't like they could screw up and give me a bad shave, right?

Bald is bald, isn't it? It's not like getting a bad haircut!

Besides, I liked the spontaneity of it.

I also liked the anonymity of it. While I had spray tanned there many times, I had never gone to a hairdresser in that salon. I'd always walked in the door, passed by the hair and nail salon, and walked upstairs to the day spa for my private tanning sessions. I had noticed the hairstylists on the main floor but never had an occasion for a conversation. I'd always slithered in and out as invisibly as possible, since my appointment involved getting completely naked and standing before a woman who sprayed my body with tanning solution. It was a little embarrassing, so I came and left almost as though, in my head, I hadn't actually been there.

Now, all of a sudden, I was going to walk in and bra-
zenly tell the women at the front desk that I wanted
to shave my head. I knew it could be perceived as a
pretty bizarre moment—totally out of character, not
that these women would even know that about me. But
one could make the leap this wasn't a typical request
for any woman walking into that salon, especially in
conservative Greenwich, Connecticut.

I had always been allowed to feel kind of anonymous
there. Okay, so perhaps they knew who I was, but we
all sort of pretended they didn't. Or at least that's how
they made me feel, which I always appreciated. And
that was why I chose to do it there. I never would have
walked into my regular hair salon down the street and
asked them to shave my head, ever!

For reasons I can't quite explain, I needed to feel
anonymous to pull it off; even then, I didn't know if
I would have the nerve to go through with it until the
very end.

Elise was more than a little shocked when she heard
my plan. I'm not sure she realized I was being serious. Now, having known me as long as she has, she
should've known I was being serious—*very* serious.
And yet at the time, I think she figured it was just talk.
The look on her face was priceless.

When Elise and I finished at the nail salon, we
walked across the bustling street in my hometown to

a wonderful little French restaurant called Méli-Mélo. We celebrated "life before cancer treatment began" by ordering two croque monsieurs, which are fancy French grilled ham and cheese sandwiches. It felt decadent and like my last hurrah of bad eating.

Afterward we walked a block to Soul Cycle, a local spinning studio (which neither of us had ever been to, by the way . . .), because someone had told me they sold bandanas that boldly read "Warrior" across the top. We found several dark gray ones with bright yellow block letters.

I bought every single one.

Does it get any better than this for my impending journey? I thought.

To me, it was just one more statement and a sign from the universe of empowerment!

And since today was all about getting into warrior mode, I was one step closer to mission accomplished!

I was doing my best to keep my life as normal as I could over the coming months. I didn't want to let cancer steal away the happy times I had planned, like the baby shower I was giving for Lindsay at my home in a month. I never even considered canceling her party when I got my diagnosis. Why would I? I only became more determined to stick with my party plans—and celebrate the joy in my life rather than wallowing in the sorrow. I had so much to look forward to. *My Lindsay was having a baby.* I was going to be a grandmother. Not that I was going to be called Grandma—uh, no. That wasn't going to work for me.

I am all about "sixty is the new forty," so I wasn't ready to wear a title that denoted gray hair. "Jo Jo" would do just fine, thank you very much!

Every time I looked at expectant Lindsay, I saw myself in her at the same age. Except that in the 1980s, when I had all three of my older girls, maternity clothes were just plain dorky-looking. No, really. When I look back at pregnancy pictures of myself, I can't believe what I had to wear on national television every day: stiff cotton A-line dresses with little blouses underneath and huge muu-muu-type dresses that made me look three times bigger than I was.

Not to mention the television cameras add ten pounds!

Today pregnant women have it so good. They get to wear tight, sexy, clingy dresses that proudly show off their baby bump. They can remain high-fashion right down to the high heels to match their slinky dresses. Believe me, I never looked as cool or chic as Lindsay: She looked incredibly beautiful, pregnant. But every time I looked at her, she brought back fond memories of me being pregnant and being on *Good Morning America.* I took all three of my older girls to work with me when they were little babies because I was breast-feeding. I would scoop them out of the crib, trying not to wake them as I changed their diapers at four A.M. and left the house in the darkness of night. The *GMA* staff was used to seeing the babies in my dressing room and always joked that during the time when I was breast-feeding, I would begin the show each morning as Joan Lunden and end the show as Dolly Parton.

There was a lot written about how I was balancing work and mothering, but I will always remember one particular morning on the program. As I usually did, I breast-fed two-month-old Jamie and put her in her crib for her morning nap while I was on the air. That morning I was interviewing some senator about President Ronald Reagan's "trickle-down" economics. You may remember that economic theory, but what I remember about it is that all of a sudden I experienced inflation and "trickle-down" firsthand. It was time for baby Jamie to feed, and my boobs knew it. Fortunately, I was wearing a silk blouse, so the cameras didn't see it, and I was able to scurry off and find a blow dryer so that, as the saying goes, the show could go on.

Those were wonderful memories.

Elise and I were ever grateful for the baby swing we kept in my office. We would wind that swing up so we could get twenty minutes of a quiet swinging baby and work on the script for the next day's show. We changed diapers while I did interviews with major magazines about balancing work and motherhood, and we always found a way to make things work.

And now my middle daughter, Lindsay, was in the same position. She would learn to juggle being a new mom while running my production company with me. I had so many hopes for her. She would be so natural at being a great, loving mom—and today very much represented the circle of life. Here we were, making plans to celebrate baby Lindsay having a baby of her own. I had to remember that life was good, even in the midst of a temporary crisis.

Before I got too far into my chemotherapy and perhaps didn't feel up to running around like Speedy Gonzalez for the big event on July 20. I was eager to get all of the shower arrangements handled while I still had great energy; I wanted to make all of the decisions and have every plan in place so the party would come off without a hitch. I wanted everything to be just right for the July 20 baby shower.

It was getting close to the time for my tanning appointment. I turned and asked Elise if she wanted to continue on with me for the rest of the afternoon, but she needed to get back home.

I drove over and parked my car in front of the Allura Salon. I was half an hour early for my scheduled spray tan. As I entered the salon, I spoke with a few of the ladies I knew at the front desk about whether I could shave my head before I went upstairs for my spray tan with Lana. The ladies all looked at me like I was from Mars.

Was I joking with them?

Had I gone crazy?

Was I about to pull a Britney Spears?

I could see that the desk manager was a little visibly shocked by my request. She fumbled over her words, saying she wasn't sure any of her stylists could buzz all of my hair off.

"But you don't understand," I said in a quiet calm voice. "My hair is all going to fall out in a week or so anyway. Isn't there someone here who can just shave it off before I go up for my tanning? This way I'll be able to get a real head-to-toe tan." I said that with a Cheshire-cat grin, doing my best to charm my way into a yes.

That was when a tall dark hairdresser named Juan emerged from the back and whispered that he would be happy to shave my head. When he said he'd do it, I turned and looked at him with utter gratitude and relief.

To be certain, there were several moments during my five-minute exchange with the women behind the counter when I internally vacillated about my intentions. But now that Juan was in front of me with such kindness and compassion, I knew I could do this.

His understanding eyes drew me to the rear of the salon. We walked to a station in the back where I calmly sat and let out a long, deep sigh. Certain no one would see us, I let him turn on his electric razor. Juan wasted no time and aimed it toward my head.

"STOP!"

I held up my hand. "Wait."

I took a deep breath and said . . .

"We need this on video, and we need some pictures."

Yeah, this was a moment I knew I wanted on camera.

I handed my iPhone to Lana, the woman who'd done my spray tans for years. I figured if I could stand in front of her naked, I shouldn't care if she saw the top of my bald head.

"Okay, Juan. Go ahead—do this before I change my mind."

I liked the sound of that. I was making the decision for myself. This wasn't being done "to me."

I'd come here by myself.

I'd asked for it by myself.

No one was holding my hand or spiriting me on.

I was in control, and in the process, I felt like I was becoming a true warrior in my battle.

As Juan moved the electric clippers closer to my head, I looked into the lens of the iPhone camera that Lana was pointing straight at me. I think I said, "Okay, here we go, it's all coming off," and I think I may have also said, "Make me G.I. Joan," because there was a part of me that needed to feel that right there and then.

It was a little scary; I'm not going to lie.

Actually, it was more weird than scary.

I don't know what I expected to see in the mirror in the moments after the final sweep of the clipper. I suppose I expected my head to be all pink with no hair, like most of the bald men I see, or maybe even a little newborn baby. But in reality, it was more like a really, *really* close buzz cut.

When Juan took ALL of my blond hair off and only a little stubble remained, it looked like I was a buzzed brunette.

I'll be damned!

My natural hair down at the roots is brown!

Oh yes, I remember now.

And let's not forget those two tiny little gray patches up around the sides of my face.

Sexy mama!

How vain am I that my immediate thought was how badly I needed a bottle of bleach poured on my head in order to turn my new brown buzz cut platinum?

Or would it be better to get some brown hair dye and cover those little gray patches up on the sides?

Or maybe I should just let it go, because theoretically, it was all supposed to fall out next week anyway?

I stared at the mirror for a minute or two and didn't say a word.

The more I looked, the more I kind of liked it.

I felt tough—like *G.I. Joan.*

There was something very empowering about this experience, and yet afterward, I slowly got out of the chair, reached down, and picked up all of my hair from the floor. I'm not exactly sure why I did it. Somehow I must have thought I was saving it for posterity. It sure wasn't going in my baby book! I placed the hair in aluminum foil and hid it in the back of my closet. There are some things we all do in life that we can't explain. For me, this is one of them.

I thanked Juan for his kindheartedness, went upstairs with Lana, and got a full HEAD-to-TOE spray tan. The reality of what I had just done didn't sink in until it was time to put my clothes back on. For a moment, I stood motionless—I barely recognized the face staring back at me in the mirror. And then I slipped my new wig on as if nothing unusual had happened and made my way downstairs to pay.

Juan was waiting for me at the register. He refused to let me pay him; he said he was proud and honored to have been a part of my metamorphosis.

Wow!

I didn't expect that kind of love or support.

Way cool!

No one could tell, but under my wig, I was now bald.

Suddenly, I was feeling quite confident and cocky, like I had donned my warrior gear and wasn't going to take it off anytime soon. I got in my car and nonchalantly went about the rest of my day, running errands. Not a single person I came into contact with had any idea what had just occurred. No one was any wiser that a bald head was lurking under that blond wig.

It had been a while since I'd checked my phone for messages. I saw that Jeff had sent a group text to the family, informing us that he had taken Max to Walmart for a haircut. The text included a picture of Max standing in front of the salon inside the store. This was a joke and a reflection of Jeff's off-the-cuff sense of humor. You see, several years back, he'd taken Max to that salon when the regular barbershop was closed. They ended up cutting off all of his fabulous long locks. I was so upset that I made Jeff swear he'd never take Max back there again. I knew him well enough to get that it was a spoof text to me and my girls. However, I was in an especially playful mood. I responded to his news with a rather unexpected answer.

I wrote back: *Oh yeah? I can top that! I just shaved my head!*

I didn't think it would cause such a commotion.

While I went about the rest of my day, my family went into a tailspin. I was in the phone store, working on getting an upgrade to my cell phone, when Sarah called from Los Angeles. "Mom, what is going on? Did you really shave your head?"

"Yup, and I have to go. I'm standing in the phone store. Love you!"

She must have thought I'd gone crazy and completely lost my mind.

Moments later, I received the same call from Lindsay.

"Honey, I need to call you back," I said.

I wasn't trying to make light of the situation. I guess I was trying to just be "normal." I didn't feel like explaining my decision to anyone. I owned my choice and wasn't about to apologize for it. Besides, what could they do about it now?

Behind the scenes, Jamie, Lindsay, and Sarah were frantically trying to reach one another, trying to figure out what was going on and who was with me.

They couldn't figure out how this possibly could have happened.

Well, it did.

By eight-thirty that night, everyone had calmed down. I sent out a text to let them know I had made a decision that I was more than my hair, and I trusted they would not love me less because I had less hair.

Jeff immediately texted back, *Actually, I love you more.*

Crazy, isn't it?

Crazy, rockin', friggin', amazingly, powerfully cool!

I had morphed into G.I. Joan.

I hoped I still felt the same way tomorrow.

Chapter 11
Creating My Battle Plan

One of the most important things you can do is remember the power of girlfriends . . . girlfriends saved my day.

JACLYN SMITH

Actress, diagnosed with breast cancer in 2002

I was desperately in search of G.I. Joan when I went to my second chemo appointment. It was my last scheduled chemo treatment in New York City before I'd move up to Maine. Lindsay planned to meet me at Dr. Oratz's office so I wouldn't be alone for the treatment. The only thing on my mind that morning was the nurse trying to stick the needle in my arm again.

Yikes!

Unfortunately, my worst fears were met as we got off to a tough start again. The first not-so-perfect needle prick hurt really bad—so much that I began to weep. I had to pull myself together and stop crying like a little girl. *I must press on. Be strong,* I thought.

Well, that's sometimes easier said than done. Eventually, the nurse found a good vein and began administering Benadryl and anti-nausea meds before beginning the Taxol and, for the first time, a dose of carboplatin. I spent the next three hours trying not to cry.

Where is G.I. Joan?

Lindsay hugged me and made contact as often as I'd let her. She could see I was in pain, physical and emotional. This wasn't the way I wanted my daughter to see me, especially my pregnant daughter. I didn't want the stress of my situation negatively impacting her in any way. Oh, how the tables had turned. I sat in my comfy recliner as the chemo chemicals dripped into my arm, closed my eyes, and remembered Lindsay as a little girl, holding her hand at the pediatrician's office, telling her everything would be okay when she had to get a shot. And now here we were. My daughter, grown up and expecting a child of her own, was now holding my hand and comforting me in the very same way.

"Press on. Be strong."

C'mon, G.I. Joan. Where are you?

When I awoke the following day, I looked in the mirror at the brown stubble atop my head and thought, *Who are you?*

Even though I knew it was probably going to fall out in the next week or two, I wanted to make my light brown stubble blond stubble. Until it fell out, I just wanted to see a blonde looking back at me in the mirror. I had some hair color at home that my colorist always packed up for me for when I was on the road, so Emir could touch up the roots on the go.

Perfect! I thought.

I was headed to New York for a final wig fitting and an appearance on Dr. Sanjay Gupta's show at CNN. When I walked into Bitz-n-Pieces, I announced, "Before we go any further, I need to tell you guys something! I shaved my head yesterday, and we need to make my brown stubble platinum. *That* will make me happy."

Emir jumped into action and got the hair color out and on my head lickety-split. Forty minutes later, I was a blonde again—well, sort of, kind of. I mean, whatever hair was on my head was blond— platinum eighties Madonna blond. When I turned and looked in the mirror, I saw . . . *me.*

Now, even though I had initially set out to do only one interview about my diagnosis, when the request came in from Sanjay Gupta, I couldn't say no. I really admire him: he is so smart, so reputable, so caring, and a terrific interviewer. During our interview, I made it a point to say that my course of treatment was what *I* decided to do with *my* doctors, but there are *many* ways for a patient to go down this path. No two cancers are alike, so you can't compare yourself to your friend, your sister, or even me. Everyone's cancer is unique. Each patient and her doctor(s) need to decide which tools are best to use for her cancer, her life, her history, and her pathology. It can be confusing to hear from people who have had different experiences, but it is because everyone is different that there are many "right" ways to treat breast cancer.

Sanjay revealed that this was the first time I was being seen on camera wearing a wig.

Okay, there it was. Reveal # 2.

That cat was now out of the bag.

I could breathe.

Surprisingly, hearing him say that wasn't as bad as I had anticipated. Amazingly, he thought I looked very much like myself. In fact, if Sanjay hadn't said something, I doubt anyone would have noticed, since I hadn't told anyone outside my husband and three oldest daughters that I'd shaved my head.

After the interview, Lindsay and I walked across the street to the wig salon and finished up my fittings. Lindsay's husband, Evan, arrived at the salon so we could all drive to Maine together.

I hopped in the backseat of their Jeep Cherokee, and I couldn't have been happier to be leaving the frenetic energy of Manhattan behind. It had been a whirlwind couple of weeks. I was exhausted from the schedule and mentally drained from the emotional toll of the experience so far. As Evan drove north, I snuggled up in the backseat and closed my eyes. We were leaving New York on a Friday afternoon, so we would have a long drive ahead of us. Frankly, I was pretty oblivious to all of the other weekend tourists hitting the road at the same time. I was in another world and delighted to be headed to our summer home. Within moments of curling up in the backseat, I was fast asleep.

While there is never a good time to get cancer, at least I could spend my summer on an astoundingly beautiful lake in southern Maine, taking my treatments and getting ready for the women's wellness camp that I run every August, Camp Reveille.

After waking up at four A.M. for nearly two decades and raising seven kids, I understood the need for a little escape and play therapy. I'd spent many years living an energetic, healthy, and lively lifestyle; logging obligatory hours at the gym; and doing everything I could to stay fit. But when I met my husband, I discovered a fitness secret that trumped anything else I'd experienced.

For ten years I'd been going to Maine to spend my summers at Camp Takajo. I'd swim, sail, hit the tennis courts, scale a climbing wall, and work out with a terrific fitness trainer. At the end of every summer, I always found myself in the best shape, my spirits lifted, highly energized, and truly inspired. It occurred to me that if I felt that way, I could give other women the same experience—at least for a weekend.

Why couldn't I plan a weekend getaway where like-minded women could share that exhilarating environment?

After reaping the benefits myself for ten summers, I got busy planning a fun-filled, joy-inducing, soul-nourishing experience for other multitasking women to spend some much-needed time concentrating on their well-being in one of the most serene, peaceful, and enjoyable havens imaginable.

I wanted women to have the chance to take up archery or arts and crafts, scale a fifty-foot climbing wall, or take a dance class. I wanted inspirational speakers to share their knowledge on everything about relationships, fashion, health, and finances. And I wanted to plan fitness classes that ranged from yoga, Pilates, core training, strength, balance cardio, to self-defense, and even kickboxing. Since the camp is on Long Lake, there had to be activities on the water and, of course, hiking among the majestic pines to invigorate the mind, body, and soul. And no weekend getaway would be complete without relaxing spa services, right?

When I started Camp Reveille nine years ago, my primary goal was to make sure everyone who came had the chance to check out for a few days in order to be their best in today's busy world. It's so important that we take care of ourselves. Not just when we are sick. "Me time" is good for the spirit, good for the soul, any time.

I was anxious to settle into our summer home on Long Lake. Actually, it went a little beyond settling in, since I was still feeling the effects of the steroids I'd been given with the last chemotherapy treatment. As soon as I got there, I was cleaning every nook and cranny and closet in the house. I wanted everything to be perfect when the lousy side effects of chemo began to hit. I also knew that I would have a houseful of guests the following weekend, and I wasn't sure how I would be feeling, so I wanted to be ahead of it—just in case.

My ultimate goal that first weekend was to set up an environment that would ensure my best success with the impending chemotherapy regimen and all of the much-talked-about crummy side effects.

My oncologist in NYC had told me there were five things that would determine how well I would deal with chemotherapy:

HOW TO BEAT THE CHEMO BLUES

1. Stay physical
2. Eat healthy
3. Get enough sleep and rest
4. Drink enough water
5. Have a strong support system

Hey, if that was the formula, I could do that!

I knew I could stay physical. I had already called Beth Bielat, the fitness trainer I work out with during my summers in Maine; from my point of view, it was business as usual. The only difference was that I made arrangements for her to come to the house every day at eight-forty-five A.M. and, come hell or high water, get me out and get me physical for forty-five minutes.

I also planned to keep the workout sessions that I had always done with Beth and a number of Takajo staff ladies and Maine friends three afternoons a week. With Beth's support, as long as I had the strength and energy, I would easily stay physical. Jeff had introduced me to Beth in 2005, and I always looked forward to my summers in Maine, when Beth got me into the best shape of my life. She is really much more than a fitness trainer. She is a black belt in karate, a Reiki master, and a LifeBreath facilitator. On days when I wouldn't be up to power walks or pumping iron, I knew she would still be my spiritual mind/body guide, keeping me positive and emotionally strong.

Now, number two on the list was eating healthy. I had been on the lookout for a nutritionist to work with throughout my battle against the cancer but, most important, during the months when I would be doing chemo treatments. From the moment I'd been diagnosed, I had been reading books on cancer and added several anti-cancer cookbooks to my expansive library. I was also getting emails and calls from a number of nutritionists who were suggesting that I work with them to stay strong and healthy while enduring my chemo sessions. Deciding on a nutritionist was my next task.

Next on the list, number three, was getting enough sleep and rest.

Uh-oh. This one really worried me.

Sleep?

Sleep had always been a problem for me, and I kept hearing rumblings about insomnia as a possible side effect of chemo.

This one concerned me more than any of the others. Sleep had been my nemesis for years, especially when I cohosted *GMA*. Just knowing I had to be up before dawn made it harder to fall asleep the night before. Well, that and having a bunch of kids awake in the house, listening to music and watching TV and talking to their friends when they should have been doing their homework. I would debate getting out of bed and going to check on them as I lay there desperately trying to nod off, but all that did was keep me awake longer. Thinking back on it still raises my stress level.

However, I was in a peaceful environment in Maine, with the most beautiful views of Long Lake. I would have no children at home for the coming two months and no real distractions; this had started in the past couple of years, when my younger children started going to camp. Having the house to myself was a luxury, and while I adore the energy of my usual lively household, there was no question that the benefits of the solitude would serve my treatment. That meant I could sleep in whenever I wanted (if I could sleep in), and I could nap during the day if I chose to.

Hmm.

Napping.

That was a novel idea.

I wouldn't be going in to the office, and everyone there would make a conscious effort to keep calls to a minimum. I wouldn't accept any new work offers for the next several months and had been required to cancel several speeches that I was booked to do because the oncologist didn't want me on commercial flights when my white blood cell count was low, not to mention the concern about how I would be feeling.

So I couldn't ask for a more restful environment, and I had lots of books lined up to read.

Given the odds, maybe sleep and rest wouldn't be so hard after all!

Okay, so that brings us to number four: drinking enough water during chemo. This was one that could prove tricky for me. I was never good at drinking enough water, and now I would need to be even more diligent about it? I remembered the chemo nurse telling me that chemo sometimes made water taste metallic.

Oh, great!

So not only do I need to drink more water, but now it's going to taste like I just bit into aluminum foil!

Yuck!

I'd have to keep thinking about how to tackle this one, because I knew it was a biggie.

Last on the list but not least important was having a strong support system. Thankfully, I had this one totally covered. I had a great support team in the form of my family and friends. Although my husband would be primarily consumed with running his camps for the next two months, I had no doubt he was there for me emotionally and whenever I needed him by my side. Jamie, Lindsay, and Sarah were making plans to come up to Maine and spend time with me whenever they could be there. And finally, I had Beth, who not only oversaw my fitness and my emotional state but had become a good friend. I felt good about the immediate close-knit support I had around me and knew it would be only bolstered by occasional visits from friends and loved ones.

Once I settled in, I realized that I should have taken my first shot of a white blood cell–boosting medicine the day after I arrived.

Oops.

I needed to get myself on a different schedule than I had been used to in previous summers so I didn't forget these important things. The injections were supposed to be self-administered.

Wait, really?

They weren't even like an EpiPen, with which I could just place the tiny thin needle on my leg and then push. These shots had a cap that came off, with a super-long needle that I was supposed to stick in my stomach!

Are you kidding?

That was never going to happen.

Thank goodness I married a guy who had a modern health center, with a doctor and six nurses right on the grounds at all times, all of whom knew how to give shots.

Yes, having nurses right next door for all of those injections would be a luxury that would make this experience a lot less painful and help make my summer go a lot easier!

By my third night in Maine, I noticed I was waking up several times in the middle of the night. I wasn't sure if it was the start of insomnia from chemo or from drinking eight to ten glasses of water a day. Remember, what goes in . . . must come out.

Despite my lack of sleep, I felt good enough to start my exercise program with Beth. I wanted to jump into the summer like I always did. My mind was focused on staying as normal as I could until normal wasn't possible. My approach was to keep myself in a mind-set that felt comfortable and familiar. I wasn't sure how my strength would hold up as I got further into my chemo treatments and the chemicals began to accumulate in my system, so I thought I would plan my workouts as if I were training for "something."

We began each morning with a half-hour power walk; then we stretched for fifteen minutes to keep my muscles supple and strong. In addition to those morning sessions, I would do my three afternoon sessions with the other ladies around camp. For years I have run this session with Beth around "camp rest hour," mostly for the wives of my husband's key employees and a few of the female counselors who are able to join us. It's something we all look forward to, and I wasn't ready or willing to give it up unless I absolutely had to. It's a lot more fun to work out with a lively group of friends, and it inspires all of us to push harder and support one another in our efforts.

Our original intention for the afternoon workouts was to offer something special to the wives of Jeff's key employees, but it turned out to mean so much more to me. It allowed me to get to know all the terrific women who have accompanied their husbands, summer after summer, for decades. They've become extremely loyal friends to Jeff's dream—to create a community with a fun, active atmosphere, while building a group of fine young men of character to send back home to their parents.

As it turned out, the women were very supportive of

me during my summer. Their love and support meant so much and made me feel comfortable at a time when I felt vulnerable. There's no explaining how grateful that kind of friendship makes you feel. I've really enjoyed getting to know these ladies well, and we have bonded into a close-knit community within the Camp Takajo family.

Chapter 12
Cancer Ain't for Sissies

You're stripped down to near zero, but most people come out the other end feeling more like themselves than ever before.

KYLIE MINOGUE

Singer/songwriter, actress, diagnosed with breast cancer in 2005

Since I had always spent my summers in Maine, it was necessary for my oncologist in New York to make arrangements for me to have the rest of my chemotherapy sessions with Dr. Tracey Weisberg, an oncologist at New England Cancer Specialists. This was a real team effort that required tremendous communication and coordination on my behalf.

Lindsay, who was now seven months pregnant and dragging a little in the hot early-July weather, was visiting. She was up first thing in the morning and ready to go with me to my first session so she could meet my new doctor, too.

We made the forty-five-minute drive to Portland, a drive I'd made hundreds of times but never for something like this. New England Cancer Specialists was located in a medical facility that was lovely. It was a larger, more "general" cancer center. Any angst I had about starting chemo with a new doctor and facility quickly faded when I saw how lovely, warm, and welcoming the staff was.

Lindsay and I spent some time with Dr. Weisberg, going over my case, and I felt completely comfortable with her. I immediately felt that I was in great hands.

Dr. Weisberg wanted to know how I was handling the effects of my chemo and what symptoms I had experienced so far.

By now you know I have a need to appear strong— as if everything is fine even when it's not. Frankly, I think I need to get over that need. But that's another story. I have a tendency to avoid highlighting any weird or uncomfortable side effects because that might make me appear—you got it—*weak* or less strong and less healthy.

Are you seeing a pattern?

I believe this is something a lot of people (especially women) do with their doctors, and it's antithetical to getting the best care.

NEWS FLASH: Your doctor needs to know what's bothering you.

I knew in my heart that I needed to be completely honest with this doctor. She wanted to help me, and the only way she could do that was if I was completely candid with her about my symptoms and side effects.

I wasn't feeling most of the things she inquired about, but I wondered: Was it Pavlovian? Once she mentioned them, would I suddenly begin to feel them because she'd put the thought into my head?

The only weird feeling I was having was my stomach. It was hard to describe. Food was becoming less appealing. Dr. Weisberg reminded me that although I didn't feel like eating, I needed to. It was important to eat well. I needed to be really careful with my food choices; she warned, "No fast food or pizza in the coming months." Thankfully, I didn't even have a taste for that kind of food.

When we finished the visit, we decided my next chemo treatment would be July third, in three days. It would be my third treatment in the twelve-week regimen.

Lindsay and I made the drive back to Portland three days later, and I will admit, I was a bit nervous.

Why?

That damn needle!

Wouldn't you know it, the first time the oncology nurse tried to find a vein, the first stick wasn't a good one. It really hurt and left a mammoth bruise. Thank God her second try was successful.

There has to be an easier, less painful way to do this, I thought.

I sat for my three-hour treatment, thinking about the upcoming weekend. My oldest daughter, Jamie, was coming to Maine with her husband, George. We would be celebrating Jamie's thirty-fourth birthday on July Fourth. Lindsay and her husband, Evan, would also be there, along with Evan's parents, Louise and Jay. While I was looking forward to celebrating Jamie's birthday, this was the weekend when my doctors had warned that I would likely lose the remnants of my hair. I imagined what it would be like, bits of my hair coming out as I handed people appetizers.

Cheese ball or hair ball?

That was more than I could conceive, which was why I'd shaved my head preemptively. I knew I had made the right decision. However, in the early weeks, I was still anxious about anyone seeing me bald. And I do mean anyone—including me. Whenever I looked at myself in the mirror, I felt half the time like I was looking at a total stranger. Maybe that's why it was so difficult to let Jeff see me at first. If I looked that strange and unrecognizable to me, what must I look like to him?

Oddly, it was that same weekend when I walked into the kitchen one morning and remembered as I was greeting Lindsay and Evan that I wasn't wearing my wig or any other head covering. Instead of panicking or running back to my bedroom, I somehow felt safe sharing my vulnerability—especially with the two of them. Maybe I figured that my daughter would be okay with it, since she was such an integral part of my treatment; and Evan is just the kind of guy you feel comfortable around no matter what. Man, I'm glad Lindsay married him. They acted that morning as if nothing were out of the ordinary, which was probably why it was so easy for me to let it go.

As a result of my chemo treatments, I was waking several times during the night and was pretty much wide-awake by four or five A.M. These quiet still mornings would prove good for journaling, something I committed to doing almost as soon as I received my diagnosis. I wanted to document my journey, if for no other reason than to have a detailed recollection of each day for my doctors. Besides, since I was awake, I wanted to do something positive with my time. I think we are more honest in the early-morning hours, before we have the chance to put up our guard. At the time, I also felt like I was in transition. I was starting to feel a responsibility to support the medicine going into my body. I wanted to rally behind the science and do whatever I could to assist in my recovery. At the same time, I didn't want to seem like or look like or feel like a cancer patient.

I was so conflicted.

There were days when I forgot I was sick at all, and then there were days when it was impossible to forget.

The weather didn't offer the picture-perfect sunny blue skies I had wanted for the holiday celebration weekend. In fact, it poured every day. Instead of sunbathing, swimming, and Jet Skiing, everyone hung out inside, doing puzzles, watching movies, and spending time together. As I looked around the house, I felt incredibly happy to see everyone so relaxed.

Plus, I was surrounded by family and friends.

Who could ask for more than that?

Unfortunately, I was also really tired from the chemo, though I didn't want anyone to know just how exhausted I was. In an effort to hide how bad I was feeling, I excused myself more than a few times, stealing away for a nap here and there.

Lindsay and Jamie wanted to help me prepare food for the group so we could make a big family-style dinner. We all piled in the car, dressed in red, white, and blue gear to head to town for supplies. We were wearing big plastic star-shaped glasses, American-flag shirts, crazy patriotic hats, and Mardi Gras beads, turning a casual trip to the supermarket into a fun rainy-day activity. I'd love to tell you we turned a lot of heads in town that day, but Fourth of July in Naples, Maine, is highly celebrated, and we fit right in!

We made salads, pasta, burgers, corn on the cob, and sliced a large watermelon for dessert. It was perfect. These weekends are the special moments in life, and especially along this journey. I will always cherish those times, because they weren't about where we were so much as whom we were with. As long as I had my family around me, I knew everything would be okay.

Chapter 13
Eat, Pray, Poop

I changed my lifestyle. I have taken what I consider poisonous things out of my life. Out of my food, out of my work, out of my social circle. Out of everything. Because I want a clean, cancerfree life. And I believe I can have that.

MELISSA ETHERIDGE

Singer/songwriter, diagnosed with breast cancer in 2004

During the summers in Maine, except for breakfast, I usually ate most of my meals in the camp dining hall. Frankly, I wouldn't see much of my husband during the weeks of summer camp if I didn't join him for camp meals.

Dr. Weisberg cautioned me that I may not feel like eating, but that it was important to remember that food was my medicine now. She said it was one of the major tools in my kit. She warned me to stay away from anything raw, especially sushi or rare meat. There was to be *no* eating from buffets or salad bars, because of the germs that linger there; my immune system would be compromised. She recommended eating healthy whole grains and, if I liked potatoes, to go for sweet potatoes rather than white potatoes. Carbs were okay as long as I chose good ones—not starchy carbs, which made eating at camp even tougher, because let's face it, kids eat a lot of starchy carbs.

So while I visited with Jeff at camp meals, I decided I would be better off having my meals at home. One positive thing about cooking for one is that you don't need to have food around to please anyone else. I could eat exactly what I wanted, when I wanted.

I had been paging through a number of anti-cancer cookbooks that I'd brought, to learn as much as I could about eating right and eating healthy during my chemo treatments. Before I left for Maine, Dr. Oratz gave me a copy of *Anticancer: A New Way of Life* by David Servan-Schreiber, MD, PhD. The book is about how Western doctors are learning the importance of food and how some foods feed cancer cells and components of other foods inhibit the growth of cancer cells.

Reading books like this taught me that there were foods I should be consuming and others I should be avoiding. That was when I decided to pick up some other books I thought might help me get a better grasp of things. *Living with Cancer* and *The Cancer Diet* were two books that included recipes to combat some of the side effects of chemo. There were recipes that would help the sores you might get in your mouth during chemo and some that catered to constipation as well as diarrhea.

One thing you might notice right away when you start chemotherapy is that it feels like some alien creature has come in and taken over operation of your digestive tract and you are just along for the ride. If you throw all caution to the wind and go for that fast-food burger, you'd better pray that you can stay near a toilet. The further you get into your chemotherapy, the more your mantra becomes "Eat, Pray, Poop." Well, if you pray hard enough, you will poop . . . eventually.

TMI?

Sorry.

As a result, you quickly learn that watching what you eat can help stave off some of the crummy side effects that chemo is ready to throw at you.

Don't you just love this conversation?

Yeah, I know, it's a shitty topic. But someone has to talk about it.

It was explained to me that the chemo not only kills off the bad cells, the cancer cells, it also kills off lots of your good cells. Chemo tends to go to fast-growing cells, including hair follicles and those inside your mouth, which causes the mouth sores; the ones inside your digestive tract (thus the digestive problems); and the ones on your skin—consequently, your skin can look dry and almost paperlike. And your skin feels different to the touch, since there's not a hair on your body!

Seriously!

Imagine how your arm would feel with no hair on it.

When you rub your hand up your arm, it's as soft as a baby's tushie!

While I thought I had a grasp on healthy eating and nutrition, I had no clue what I should be eating for my health during treatment. I asked my assistant Elaine to look into several nutritionists who had been recommended to me since my diagnosis. All of them were well known in their field, though some were more recognized and successful than others. She called the various offices to find out how they would work with me and what their rates were, whether they accepted insurance, and how often we would need to meet. After all, I was in Maine and they were scattered throughout the country.

The week before I was diagnosed with cancer, I had been at a hair salon in Connecticut and met a PR rep for a physician and clinical nutritionist from Westport named Dr. Robert Zembroski, who specialized in advising patients going through chemotherapy or autoimmune problems. The publicist told me that Dr. Zembroski had written a book about his own cancer experience called *Rebuild with Dr. Z's Body Composition Diet*. The publicist said that she would love for me to meet him. Keep in mind, this was *before* I knew that I had cancer. I told her that I would be making an appearance later in the week at our local library to interview Adam Braun, founder of Pencils of Promise, a for-purpose organization that builds schools, trains teachers, and funds scholarships in the developing world. He had just published his first book, *Promise of a Pencil*. Adam and his siblings had gone to school with my daughters, and he had asked me to moderate the event for him.

Well, the universe has a way of delivering what we need exactly when we need it, because Dr. Zembroski turned up that night. He sat in the front row and introduced himself after the presentation and handed me a copy of his book. Once I went public with my diagnosis, he made it a point to stay in touch, graciously offering to work with me and share his incredible knowledge of nutrition throughout my cancer battle.

Once I was diagnosed I read his book and learned so much. He rebuilt himself after dealing with his own diagnosis and journey through cancer with non-Hodgkins lymphoma. He is considered a miracle cancer survivor because of the approach that he took in his own battle.

Dr. Z draws on the latest research and over twenty-one years of working with patients, so when the time came to find the right doctor to add to my team, I knew in my gut that Dr. Z was the guy for me.

I scheduled my first call with him right after the holiday weekend with my family. Not only was his office close to my home in Connecticut, it turned out that he had roots in Maine, too. His brother is a doctor in Augusta, and they share a number of patients, so Dr. Z visits Maine with some regularity. I often look for signs that I've made the right decision, so from my point of view, this was meant to be!

The first thing Dr. Z asked me in our initial call was whether or not I'd gotten a port. A port is a small medical appliance installed beneath the skin, usually in the upper chest area, just below the clavicle or collarbone. A catheter connects the port to a vein. Under the skin, the port has a septum through which drugs can be injected and blood samples can be drawn many times, usually with less discomfort for the patient than a more typical needle stick.

Dr. Z warned me that I could ruin the veins in my arms if I continued with the IVs. To be candid, as much as I hated the IVs, I had been reluctant to do the port procedure, as it required minor surgery. Besides, none of my doctors had encouraged me to get one, so I didn't see the rush or need to have the surgery until Dr. Z explained to me that: "once the port is in, they simply disinfect the skin and put the needle into the port. The tube goes under the skin and into a major artery in your neck. When chemo drugs go into the port and thus into this major artery, they go directly to the heart and are immediately dispersed to your entire bloodstream. It's much more efficient than going in through veins in the arms. Call your doctors tomorrow morning and schedule an appointment to get a port ASAP."

He wanted me to know there are tactics to make it through chemo in the best way possible and, maybe even more important, assure that it doesn't do any permanent damage to your body. He warned me that the strong chemo drugs can cause damage to the kidneys.

Our conversation turned to nutrition. He explained that chemo is cumulative. So often people think, *I'm doing great with this! I'll be fine.* But as anemia builds and your white blood cell count drops, you start to feel worse.

He was quick to say he didn't want to scare me off, but what he was about to propose would feel extreme. He was going to basically take me off *sugar, wheat, and dairy.*

Whoa.

First, that wasn't exactly what I had been hearing from my oncologist.

Second, what was I going to eat?

Dr. Z explained that I was going to start a *clean* eating plan. In the simplest of terms, clean eating is about eating whole foods or "real" foods—those that are un- or minimally processed, refined, and handled, making them as close to their natural form as possible.

Okay. I could do that. Instead of scaring me, it fueled me to clean out my food pantry and fridge and throw away tons of food filled with wheat, sugar, and dairy. It felt good, cleansing, like I was once again gearing up and finding my Warrior Mode.

Too bad we have to get hit in the head with a two-by-four to really be serious about our health. But I had an amazing incentive to do it right this time. Although I didn't know it, I had been paying lip service to label-reading and eating clean for way too long. Now the time to follow through was here—BIG-TIME!

This was my second chance.

I doubted there would be a third.

Okay, so we were going to attack these cancer cells with FOOD. Dr. Z assured me that some of the best medicines you can find to fight cancer are the *right foods.*

Absolutely NO WHEAT, NO DAIRY, AND NO SUGAR.

That meant no whole-wheat anything and no or minimal grains.

He suggested substituting black rice or brown rice. Brown rice pasta, brown rice breads: two things I had never tried.

He also warned me to stay away from fatty red meats and all things fried. My best bet was to stick with lean chicken and broiled fish. I could eat all flat fishes as long as they were cooked well. Oh, and in case I was wondering, sushi was completely off the menu.

Wow!

That was our *starter* conversation.

Dr. Z said he was coming to Maine in a few days and would love to have another session.

It would take me a few days to wrap my head around everything he had thrown at me.

I told him I was completely committed and wanted to take this new clean-eating plan to the next level. I was ready to do whatever was necessary to aid and abet my health.

When he came to Maine, Dr. Z and I decided to make a trip to the grocery store together to do my very first official "no wheat, no dairy, no sugar" food shop. What an eye-opening experience that was! I felt like an alien who had landed in the middle of Whole Foods for the first time.

While I had been to Whole Foods many times in the past, this was a totally different experience. Dr. Z walked me up and down the aisles, showing me foods and explaining ingredients I never knew existed. I wrote my first cookbook in 1994, *Joan Lunden's Healthy Cooking,* and I'd jokingly asked, "What is quinoa, anyway?" Now here I was, buying it! (I still don't know exactly what it is or where it comes from, but it's pretty tasty.)

When we finished shopping, I came home and emptied our food pantry of all of the Tostitos and other chips, cookies, and crackers that didn't come home from my "healthy shop." I threw away all of the bread in the house, even the whole-wheat English muffins I once held on to as my healthy breakfast choice, and everything that didn't fall under my new guidelines to healthy eating but might have still tempted me. I wasn't taking any chances with my eating or my health going forward.

I was the most concerned that I'd miss bread and my morning English muffins; however, I acquired a taste for brown rice bread with a smear of Tofutti cream cheese and a tiny bit of raspberry preserves on top. Surprisingly, I found I didn't miss any of the real things a bit—especially the cream cheese.

I found the transition from cow's milk to almond milk very easy—frankly, I can't even tell the difference. I also tried rice milk. I liked them both, though rice milk tended to be a little watery for my taste and wasn't as good with steel-cut oats or in coffee. So for my coffee, I started using a coconut-milk creamer, which is a totally delicious substitute for regular milk.

Instead of regular pasta, I substituted brown rice pasta, which took a little getting used to, and chose black bean chips to dip into my beloved salsa and guacamole. After my semester at sea, I spent two years of college studying in Mexico City, which left me with a love of the food, that's for sure. I was happy to find a satisfying substitute chip so I didn't have to give up one of my favorite snacks. Although I am more of a salt than a sugar lover, I'll now occasionally indulge in Tofutti ice cream bars, which aren't quite as delicious as the real deal, but they did satisfy my craving for something sweet and cool for dessert whenever I got the urge, which wasn't often.

What I really discovered during my first shopping trip with Dr. Z was that there are a lot of good substitutes and options, but it would take some time and trial and error to find the ones that worked for me. I also realized, perhaps for the first time, that while I'd thought I had a handle on what eating healthy was all about, I had been fooling myself for years.

I was so consumed with counting calories and watching my carb intake that I never paid attention to all the other crap I was ingesting along the way. I never spent a moment reading the labels on the food I was eating or feeding to my family. When I began to take notice, it scared me to think we were eating chemicals with names I couldn't pronounce. Surely this wasn't the way our bodies were meant to operate. In a way, I felt so fooled. If something said "healthy" on the label, I bought in to it—hook, line, and sinker. I never bothered to read the back of the box, jar, or can to check out what was inside.

This was an epiphany of colossal proportion.

An awakening I hadn't seen coming.

I knew this was the beginning of a different journey for me—one I was committed to in a whole new way, from a completely different point of view.

Why?

While I always knew my health depended on my nutrition, this time my life was at stake.

Oh yeah.

That'll rattle your chain.

If getting cancer isn't a good enough reason to get hold of your eating and nutrition, I don't know that anything will get you there.

I had always been a woman who thought I walked my talk.

I put up a good front.

I talked about it—even wrote about it.

And now I had to change my path.

Big-time.

Chapter 14
Port of Entry

We never hid anything from the kids. I feel whole again. I really do. I told them Mommy's boo-boo is better now.

WANDA SYKES

Actress, comedian, diagnosed with breast cancer in 2011

After talking to Dr. Z and doing quite a bit of research, I decided that getting a port made a lot of sense. I will confess that I grappled with the idea of putting myself through a medical procedure—did I really need to go into a hospital and do this?—but after weighing out the pain I'd endured during my first two chemo treatments, then reviewing a lot of documentation from other patients who had written to me about their own experiences, coupled with what Dr. Z told me, I felt like it was a no-brainer. I couldn't imagine why I would choose to suffer more than necessary, let alone create unintentional long-term damage to my veins. In the scheme of things, I was hoping my chemo treatments would be a blip on the screen. I didn't want to live with the fallout of those twelve sessions longer than I had to.

It was my body, my decision.

As I was learning, this journey was all about choosing what was right and best for me.

Give me the port!

Jeff drove me to the Maine Medical Center, where Dr. Chris Baker performed the one-hour surgical procedure to implant a power port in my chest. I was awake, in what's called a "conscious sedation" state. I had an issue with the "conscious" part of that description at first, but it's not as bad as it sounds. The anesthesiologist said, "You will be awake and you will hear us, but you totally won't care."

And he was absolutely right.

Members of the interventional radiology team at Maine Medical chitchatted with me throughout the procedure as Dr. Baker made a small incision in my chest using images from the huge MRI machine that hovered over my torso. He inserted a small tube into my major artery. A few inches below that, he made another incision and slipped the small port under my skin. Once the two were hooked together, Dr. Baker stitched me up, and I was good to go.

When the procedure was over, they left a needle in my chest for my next chemo treatment, which was scheduled for the following morning.

While that is quite normal, the idea of rolling over in bed with the needle sticking out of me made me a little queasy. I was very groggy and sore, leaving the hospital, but so happy to have it over with, knowing chemo would be a little easier going forward.

The next morning, my mother-in-law took me for my chemo treatment. It was the first alone time I'd spent with Janey since my diagnosis. Janey was all too familiar with the breast cancer journey; many years ago, she'd had a lumpectomy and six weeks of radiation. All of the women in Janey's family were on high alert, as her elderly mother had also had breast cancer. Knowing what Janey and her family had been through, I was well aware of how fragile life can be and how quickly things can change.

Buddha said, "The trouble is, you think you have time."

There is a reason so many people buy in to the Buddhist wisdom.

We all think we have time.

But do we . . . *really*?

I know I'm not supposed to go there, but to never ponder this question, especially now, would be denying that I am human.

One day after I wrote a blog entry, I received a message via my website that really touched my heart. It also really got me thinking.

I am glad that you are doing well with your treatment; you just keep fighting. My daughter was diagnosed with triple negative breast while pregnant, she delivered her baby, a miracle, and she continued to fight for eleven months. Unfortunately she lost her battle this July, leaving a husband and two little ones. Her husband set up a fund in her name to help support TNBC research and to help children that have lost a parent to this disease.

—MM

After reading this, I sat for a moment or two, completely stunned, frozen in my seat. I read the message a second time, and afterward I cried. For the first time since being diagnosed, I thought about the reality that I, too, could die from this disease.

Why hadn't I honestly considered that consequence before?

I guess it's because I'm a positive thinker and a natural-born fighter who doesn't consider losing a possibility.

And I still believe that in my heart of hearts.

There wasn't one ounce of me that didn't believe in my ability to stay strong and fight, couple that with the amazing modern medicine available to us, and beat my disease.

But . . . it *is* cancer.

What if it spread and got the best of me?

Didn't I have to at least give that outcome *some* thought?

Didn't I have to consider the "what if"?

It was a totally scary thought.

Not so much about what happened to those of us with cancer—I mean, in the end, did that matter?

We'd be dead.

It was really about those we leave behind.

In my case, that's Max, Kate, Kim, and Jack, who are still so young.

And what would Jeff do?

Yes, he is the most capable, organized individual I know, but come on. We are so connected, the two of us, and we live our lives as a family unit—planning and enjoying every minute of it, every different phase of it.

We savor it.

And what about my older girls, who are just beginning their adult lives and will have so much to share and so many questions to ask me?

Lindsay was having her first baby with Evan, and I was so excited for her. The idea of them not having me for these monumental life experiences was more than I could bear to think about.

So the thought of all that brought tears—real tears, which I hadn't yet encountered during this journey. I'd been fighting and staying positive because that was what I thought I needed to do to ensure a good outcome. Surprisingly, I hadn't cried much.

Not yet.

Not until that moment.

I'd heard other women talk about how much they cried.

Why were they so consumed with crying and sadness about their cancer and I was not?

Was it because I was so cocksure I wouldn't succumb to it?

Hey, I was also really cocksure I wouldn't ever get it, wasn't I?

But I got it.

Damn.

The concept of death had become a reality to me when my brother died eight years before, and then a year before, when I lost my beloved mom.

I had experienced other deaths in my life, tragic deaths of friends, and of course my dad—but I was so young then.

In recent years, I had experienced the finality of it all.

When my dad died, it was impossible for me to grasp the concept of death because of my age. Although I understood that his plane had crashed, I made up my own story as a means of coping with my sadness and loss—my own "truth" about what happened in that plane crash. I remember our home being filled with people who had come to pay their respects. I overheard someone telling my mom that at my father's funeral, there would have to be a closed casket because the plane had crashed at full speed and no bodies were ever recovered.

Well, that didn't happen in my version.

In my young suffering mind, my dad walked away from the crash. He was hurt but alive; however, he had amnesia. He walked until he found a house where they cared for him, but because he didn't know who he was, he could never make his way back to our family. My reality left the possibility of a happy ending, that one day he might regain his memory and come back to us. I know it's implausible; however, childhood fantasies don't always make sense.

Then the cruel finality of death hit me when my brother, Jeff, died of complications from type 2 diabetes. After he was gone, I tried to recall something about our youth, our family, or our friends. I would reach for the phone to reminisce but then quickly remember it was no longer possible. It was a very lonely feeling, because after my brother's death, my mom began to experience dementia more and more. She couldn't remember details of our life, which really made me sad. Jeff had been my only connection to our family story. With no grandparents, aunts, or uncles alive to contact, I felt like our family history would never be tapped into again unless I remembered it.

Oh, God, please don't let my memory start to go, too.

In 2013 my mom passed away at ninety-three. So now there was just me left. Oh, how I wish I still had that wonderful opportunity to just pick up the phone and say "Hi" and "I love you." I still have many moments when I want to ask my family members something that only they would know the answer to, or to share a piece of exciting news with them, but then I realize they are no longer here.

And that is what it would be like for Jamie, Lindsay, Sarah, Jeff, Kate, Max, Kim, and Jack if I died.

Allowing them to feel that way was unimaginable to me.

Unable to shake the thought of what would happen should this cancer actually do me in, I began pondering what my family would find when and if they ever had to go through my things.

Oh my God.

They are all going to think I was a totally unorganized slob!

I emptied my underwear drawer and folded each pair of panties and neatly stacked my bras according to color. Then I dumped out all of my socks and paired each and every one and folded them in neat stacks of pairs. I weeded out my closet, trying to make it look as neat as I could. Then I began opening file drawers in my home office and pulling out documents and putting them in newer, neater folders. I rearranged the folders in an order that I thought would make more sense to someone looking for important documents. A few times I found folders empty of important documents.

Why had I taken that document out of its folder?

Where was it?

I began searching through my in-box and out-box.

I couldn't find some of them, and that was making me anxious.

No matter how much organizing I did, it seemed like I only scratched the surface that day. There was so much left to do to make things more orderly. I opened another drawer.

Oh no.

It was full of thousands of photos that hadn't been filed.

That task would take me forever.

I didn't have forever.

I could be there for days—no, weeks—months!

Putting all of my affairs into better, neater order suddenly seemed like a priority.

Did I really need to go to this effort now?

It had become so overwhelming.

Wait, why was I making myself crazy?

What happened to that positive attitude that said, "I will beat this!"

I needed to go back there.

At least I wouldn't have so much work to do.

But in a silly, stupid kind of way, my "mortality mayhem" was good for a little run. I got a lot of drawers cleaned out.

However, I was exhausted—mentally and physically.

I had to let go of the thought that I might not make it through. I needed to return to my "glass half full" or "You bet your sweet ass I'm going to win this battle" attitude pronto.

I had a tennis lesson scheduled at nine A.M., and by gosh, I was going to it!

But before hitting the court, I wanted to hit the beach to grab a little sun. For the first time since shaving my head, I was going to take off my bandana, get outside, and get some sun on my pinkish-white bald head. My mom used to tell me that tan fat looks better than white fat. If that is true, then a tan bald head could be kind of cool, right?

Or would that just apply to men?

Yeah, I'm pretty sure that's just on men.

But I thought I could pull it off.

This brought up one last question that made me laugh every time I got into the shower.

What was I supposed to use to wash my head?

Shampoo or body wash?

Chapter 15
You're Never Too Sick to Throw a Party!

You get pushed down, you pick yourself up and move on. I've never been a victim.

SHARON OSBOURNE

TV personality, diagnosed with breast cancer in 2002

It was now the middle of July, and I was scheduled to have my fifth chemo treatment. As soon as that was done, I planned to head back to my home in Connecticut for Lindsay's baby shower. I couldn't wait to get all of her girlfriends together, many of whom were new moms themselves, to celebrate Lindsay's pregnancy. I had also invited some of my friends who had watched

Lindsay grow into a beautiful young woman and who were excited to share in the joy of her becoming a mommy. And what a mommy she would be! Lindsay has always been incredible around children; whenever she walks into our house, she lights up, drop her bags, and runs to scoop up the twins to play with them for hours. I could only imagine how she would be with her own baby. I could hardly wait to see her as a mommy.

Worried that I might not feel up to the task of throwing a party for forty people, Lindsay graciously did her best to talk me out of hosting the event, but there was no way I would cancel. I wanted to do this. I've never been too sick to throw a good party! *Especially* a baby shower for my first grandchild!

By this time, I wasn't doing any long-distance driving alone. It's not that I wasn't able to drive. It was more that everyone agreed I shouldn't because I was tired, and my physical reaction to the chemo was still unpredictable. Knowing this, I reluctantly called Jamie, who lives in New York City, and proposed that I get someone from camp to drive me halfway down, at which point perhaps she and George could meet me. Jamie would have no part of that plan, saying she and George would drive up to Maine together, stay overnight, take me to my treatment the next day, and then drive me back to Connecticut. I felt so guilty having them go through all of that. They're both incredibly busy executives, Jamie in PR and George in the music business. The hectic busy life those two lead would make most people dizzy; I really didn't want to burden them.

It's been challenging for me to graciously accept these types of gestures because I've been so independent my entire life. I've never had to ask for help, let alone depend on anyone, especially my children, for anything. Some parents look forward to the day when their kids will take care of them the way they took care of their kids, but not me. All that did was make me feel old and incapable, which I didn't see myself as—and that only made it harder for me to swallow my pride, smile, and say, "Thanks." Jamie was actually happy to have the opportunity to be with me for one of my treatments, to meet my doctor whom I'd told her so much about, and to have a chance to ask questions.

The appointment was over around one P.M., and we were all hungry. We asked the nurse, Jenny, if there was any place nearby where we could get something healthy to eat. Jenny knew about the clean-eating program I'd been following and suggested a nearby health food store. Jamie was excited about this adventure, going on a "healthy food shopping spree" with me. Jamie has been a healthy eater for years, and George looks ten years younger than his age—he's a runner, he works out like a maniac, and he is what I'd call an ultimate clean eater. (Oh yes, and he's worked in the music industry his whole life, which also keeps you rockin' a younger look.) When those two got together, they took healthy eating to a whole new level. Even their Yorkie, Stella, had to become a healthier eater—much to her dismay. No more table scraps for poor Stella!

Not only are George and Jamie health advocates, but they are out of bed before the sun comes up and in the gym at the crack of dawn. To them, an eight A.M. meeting is a late start. I am in awe of their enthusiasm and endurance. They are constantly running 5Ks and signing up our whole family to run in shorter races with them. I love it and think it's a great example for my younger ones.

For years, Jamie had been trying to get me to "come over to the other side" and eat more like they do. Now that I was on Dr. Z's anti-cancer clean-eating plan, she couldn't wait to walk the aisles of the health food store with me. It was a fun bonding moment for the three of us: We were like three kids in a candy store, albeit a sugar-free, dairy-free, gluten-free one! We filled our carts with veggie dishes, healthy veggie chips, and lots of water. When we were done, we piled into their car for the ride back down to Greenwich. I'm not sure I ate very much before falling asleep in their backseat. The chemo and the anti-nausea meds had a tendency to do that to me.

My daughter Sarah would be arriving from Los Angeles later that night and staying with me at the house. I was really looking forward to some mommy/daughter time over the weekend. She'd been living in L.A. for some time, working in television production and loving the California sunshine (which I couldn't argue with, since I'm an original California sunshine girl), and I missed seeing her on a regular basis. She was planning to move back home in a month or so to help run my company once Lindsay went on maternity leave, which was truly exciting, but I couldn't wait to see her over the weekend. This would be a super-fun sleepover, movie-watching, crossword-puzzle-solving marathon. Just the way we always loved it! And Sarah would help me with all the last-minute shower plans.

With a baby on the way, Lindsay and Evan decided they needed a larger apartment in New York City. As it turned out, that weekend was the same one they would move to their new place. Normally, I would be there helping with the moving effort, but I simply wasn't up to that task. While I wanted to pitch in, I didn't have the energy or oomph. Besides, it was taking everything I had to pull together the shower. That was my first priority.

When I got back to Greenwich that night, it felt a little strange to be there without Jeff and the younger kids. Our home is usually so chaotic, and that night it was oddly quiet and calm. While it was lovely to have my older girls around, it was a completely different vibe from what I was used to in that house. Jamie and George stayed the night, along with Sarah, who had settled in after a long flight.

As my girls have gotten older, they've all gone on to live very independent lives. (Gee, I wonder whom they get this from.) Each of them has grown into an amazing, incredible, and accomplished woman, making an individual mark in the world. With Sarah in L.A. and Jamie and Lindsay in New York, we don't all get together as often as we once did. I can't explain why I felt sad that night, why I thought that it took a crisis to pull our family together, when in fact, we had gathered for a celebration of life: a brand-new life we were about to welcome into our fold. The more I thought about it, the more I realized how blessed I was to have these three amazing women in my life, and I got to call them my daughters.

The next morning was my only day to get as much done as possible before the shower. Sarah and I set off with our to-do list, which included picking up flowers; confirming delivery times for the tables, chairs, linens, catering, and cake; and hitting the liquor store. By the end of the day, I noticed that I had raised hard sores on my forearms. They weren't there when I left the house earlier.

What the heck could they be?

They didn't really hurt, but they sure looked nasty. My best guess was they had been exposed to the sun all day as I drove around town. Like a bonehead, I hadn't put on sunscreen because I didn't think I'd be "out in the sun."

Okay.

So I guess I was now super-hyper-sensitive to the sun, just like my doctor said I would be.

Note to self: Be more careful going forward. And wear sunscreen at all times!

The next day, for reasons I can't explain, I felt like I was flying high. I don't know if it was pure adrenaline or the bag of steroids they gave me before my chemo, but I was like Wonder Woman that day.

No, maybe I was more like the Bionic Woman.

I had all of the tables set before anyone else in the house was awake. It could have been nervous energy. Whatever it was, for a while there, I felt like a million bucks.

Having everything done early gave me time to take my daughters out for brunch at a wonderful little French restaurant famous for its fresh pure ingredients.

I was working hard at staying the course with my clean-eating plan. I carefully looked through the offerings, starting with the list of soups, which felt the safest to me. When the waiter came to our table, I asked about every single ingredient in each soup: "Is there any cream added? Any sugar? What's the stock base?" When he told me the one I wanted was made from pure butternut squash and pureed roasted apples, I stopped him. "Perfect," I said.

I perused the menu for my main course, and while there were lots of dishes that looked good, it seemed like each had at least one ingredient that was not on my approved food list. So I'd need to modify a dish or two a bit—which wasn't difficult to do. *The beet salad over arugula with goat cheese could work if they remove the cheese and put the dressing on the side,* I thought.

This exercise proved I could eat out without a lot of distress or deprivation. It wasn't all that hard, and the waiter was happy to oblige my requests.

Unfortunately, as the day passed, I found myself getting terribly fatigued. I felt nauseated, and I needed to go home, get into bed, and rest. I was very worried that I might feel this way for the shower, when I so desperately wanted to be bright and happy for Lindsay's big day. I was beside myself when I broke the news to the girls that I needed to cut our day short. Of course, they understood, but it wasn't how I wanted the day to end.

When I awoke the following morning, I still wasn't feeling great. Lindsay and Evan had moved over the weekend; she had to be exhausted from all of the unpacking. I was more worried about her than I was about me. I definitely didn't want her to know how crummy I was feeling. I wanted to rally—to put on my best game face and get through the day as though nothing were wrong.

I rested until the caterers arrived at three in the afternoon. By then, I was feeling a bit like my old self and was able to call the shots. By the time our first guests arrived, I was feeling great. Every time a guest walked in the door, one of my older daughters quickly announced that "Mom is only taking air kisses, no touching or hugging or kissing," because they were concerned about my low white blood cell count and were being extra-careful not to expose me to other people's germs.

Toward the end of the shower, Lindsay gave a very emotional speech that brought the entire room to tears. It wasn't a prepared speech. It came straight from her heart. I smiled because she had become such a gorgeous woman, and now she was about to become a mother, too. My heart was full, truly, listening to her and realizing that my baby was having a baby.

A couple of days after the excitement of the festivities had faded, Sarah and I headed north to Maine, but not before lending Lindsay a hand with unpacking all of her baby gifts and getting settled into her new apartment. None of my girls had ever moved into a new apartment without me helping them place their art on the walls—make that *my* art on their walls—and I wasn't willing to forgo the tradition. As soon as the last picture was in place, Sarah and I hit the road. About two hours into the drive, I could see that Sarah was sleepy, so I decided to take the wheel for the rest of the six-hour trip. I felt good to be taking care of her for a brief time.

I was having a rare and especially good day, full of energy and positivity. There was a lot to look forward to. Camp Reveille was three weeks away, and I was about to hit the halfway mark in my first round of chemo. Meeting that six-week milestone felt like a big accomplishment. That was the mind-set that would help me get through this unplanned journey. Early on, I made the decision that my attitude would dictate my experience, and my experience was absolutely dependent on my attitude. I looked at every small step as a tiny success leading me toward my ultimate bigger goal: to beat the crap out of this disease. That was the only way I could remain optimistic and happy.

Hey, I always try to look at the glass as half full. My mom taught me this wise life lesson many times throughout her life. Glady was my personal guru of positive thinking. She could be angry with me and scold me, but fifteen minutes later, she had let it go and was at my bedroom door with a big smile, ready to move on.

I know it couldn't have been easy for my mom to always stay so positive. When I was thirteen, we were living an idyllic life in Northern California. Though my dad was a cancer surgeon who spoke at many cancer conferences, he was also a businessman, building a medical office building and a community hospital.

After my father's untimely death, I watched as my then-forty-one-year-old mom picked up the pieces of her shattered life and figured out how we would go on. She had to make heads or tails of our family finances while raising two young teens alone. She transitioned from being a stay-at-home mom to a working mom. So I learned at a young age how fast life can change. Sometimes it really does happen in a blink of an eye.

Living through that experience and watching my mom persevere through the toughest of times and still find a way to smile had taught me the importance and value of a positive outlook on life—something I have tried to instill in my children, too.

Now more than ever, I knew my attitude would play a big role in how I dealt with my cancer treatment. Staying positive, especially in front of my children, was critical.

Sarah took me to my sixth and "halfway there" treatment. This was her first time accompanying me to one of my chemo sessions. I had come to really adore Dr. Weisberg. She was so patient and thorough every time I saw her. She diligently went through all of the questions she asked at every visit, asking whether I'd felt tingling in my fingers or toes.

NO.

Had I felt my heart racing or skipping a beat?

I said NO, but the truth was, I might have.

Why wasn't I being totally honest?

At this point in my treatment, I needed to trust in the process, but for some reason, I was still hedging on my answers, seemingly so I would appear strong and "fine." Mainly, I wanted to be more fine.

Dr. Weisberg asked if I had experienced any headaches or nausea.

I told her there had been no headaches, but I'd had some bouts of nausea, and my belly did feel as though it had been taken over by aliens who were fully in charge. My belly was perpetually bloated because my digestive tract couldn't do its job anymore; I was in a constant state of discomfort and always on the brink of heart-burn. I told Dr. Weisberg that I had been using the natural laxative Senokot and the stool softener Colace, which her nurses had told me to use. Dr. Z had also suggested that I take an aloe pill at night, then a swig of aloe juice in the morning before eating, and another in the afternoon on an empty stomach. Dr. Weisberg thought that was a terrific alternative way to help without putting any more chemicals in my body. As long as I got results, it was all fine with her.

I also went over some of the vitamin supplements that Dr. Z had suggested: They included B12 and C and D, which Dr. Weisberg approved of, too. It felt reassuring that Dr. Z and Dr. Weisberg were on the same page. It also was nice to know that I could present a doctor with the other's plan for my health and wellness and have them agree that it was the right course of action for my well-being.

Dr. Weisberg made a point of saying that after I'd had the equivalent of a chemical bomb dropped on me, I was managing to build myself back up by doing all the right things, and she thought that was awesome. This was just the kind of encouragement I wanted and needed from my doctor to keep me motivated with the eating plan that Dr. Z had outlined for me, working out with Beth, and maintaining my positive attitude.

I will admit, there were a lot of people out there rooting for me, and that helped me stay confident about my prognosis. Nancy Brinker, who started the phenomenally successful Susan G. Komen breast cancer organization, called soon after I went public simply to check on me and see how I was doing. I had interviewed Nancy over the years and had always admired how she held to a promise to her dying sister, Susan, who lost her battle with breast cancer at thirty-six, and how Nancy changed awareness of breast cancer in America. To receive a call from her said a lot about why she had been so successful. I was also incredibly heartened when Leonard Lauder of the Estée Lauder cosmetics empire contacted me to check on how I was feeling and to express how proud he was of the leadership role I was taking in the fight against breast cancer. His wife, Evelyn, had survived breast cancer years earlier and started the Breast Cancer Research Foundation. Evelyn died in 2011 of ovarian cancer, but Leonard Lauder jumped in and has continued to build BCRF. Since BCRF was founded in 1993, every single major breakthrough in breast cancer prevention, treatment, and survivorship has had BCRF funding. Currently, the foundation is the largest private funder of breast cancer research in the world. Leonard is truly a hero to me.

I also got a call from Dr. Susan Love, the renowned cancer surgeon who had told me during that fateful interview years earlier to follow my mammogram with an ultrasound. I told her that I attributed to her the credit for my being able to find the cancer in time to treat it. These powerhouse people have all kept in close touch with me throughout the course of my treatment.

Frankly, I have been awestruck by the power and sense of loyalty that the breast cancer sorority has, every female member instinctually reaching out and checking on other women, lending advice or simply strength. It's not a sorority that you want to join, that's for sure, but you can't argue with its heart and its passion to help other members through their battle and to ultimately find a cure.

People magazine had been in touch since I made my announcement earlier in the summer. They wanted to interview me then, but at the time, I didn't feel like there was a lot to share. I needed to place my full attention on my treatment and recovery. We agreed to circle back around later in the summer, and I'd give them an exclusive for a cover story in September. When we made this arrangement, I agreed to allow a reporter and photographer to come to Camp Reveille. However, as the date approached, I wasn't so sure I had made the right decision. I was worried that I wouldn't be feeling up to the weekend retreat.

Okay, let me clarify this. While I thought I was doing fine, Dr. Weisberg had her concerns.

When I went for my ninth chemo treatment, the nurses took my blood and we waited for the results, as we did before every session. When Dr. Weisberg came into the room to run through her familiar list of questions about how I'd been doing and feeling, I was extremely upbeat and positive, telling her that I'd been working out and playing tennis every day.

She looked at me with a bit of a confused expression and said, "Well, there's a huge disparity between what I am hearing from you and what I am seeing on your chart. Your blood work says that your white blood cell count is down to 2.7, your hematocrit [another measure of red blood cell count] is at 27.0, your hemoglobin is down to 8.6, and your platelets are down, too. In fact, they're very low, Joan."

The normal white blood cell range for my age is 3.8 to 10.6; for hematocrit it is 36–42. The normal range for hemoglobin is 11.9 to 16.0. Clearly, I was way off.

Now I was confused—and worried. I was very tempered by hearing how low all my numbers were and the realization that I was much weaker than I perceived, so weak that the doctor said she might not give me the treatment. I didn't want to prolong the course of my treatment, and quite honestly, I didn't want to "fail." I wanted to be one of those warrior patients who never missed a treatment!

Dr. Weisberg said we had three choices:

1. She wouldn't give me a treatment that week, which would give my body a chance to recover.

2. I could blast through, choosing to go ahead with my treatment, but I might have to skip the following week, since I might not have the energy I needed to be at the helm of Camp Reveille.

3. She could give me a reduced treatment, choosing to go with 120 milligrams of Taxol rather than 150 milligrams. That way I wouldn't be off my schedule by a week, but I wouldn't get even more physically slammed when my blood counts were so low.

There was no chance I'd skip a treatment. I wasn't prolonging this process unless I absolutely had to. And for sure, I needed to be on my A game and present for the ladies making the trek to Camp Reveille, which was happening in a few weeks. So many of them were fighters and survivors of their own battle against breast cancer. I needed to be there, to be strong, to be a role model and a beacon of hope.

Yeah.

It was a pretty easy decision.

I'd take door number three.

Chapter 16
Calling All Campers

With over three million women battling breast cancer today, everywhere you turn there is a mother, daughter, sister, or friend who has been affected by breast cancer.

BETSEY JOHNSON

Fashion designer, diagnosed with breast cancer in 2000

The last day of camp is always filled with mixed emotions for everyone. There's laughter and tears and lots of hugs to go around. It marks the end of another memorable summer for the boys, most of whom board a bevy of buses and head home to their anxious parents, who haven't seen them since visiting day. However, some of the boys stick around and are joined by their dads for Jeff's annual father/son weekend, which has long been a favorite for dads and sons alike. As a father, Jeff feels strongly that our children need our presence more than our presents, and there is no greater impact on a son than spending quality time with his dad. The dads play basketball and soccer and water-ski and scale the climbing wall with their sons. They become a bunch of weekend warriors, killing themselves on the courts and the fields, and then sleep on a child's single bunk bed near the other snoring dads for a couple of unforgettable nights of bonding. Ah yes, good sports they are!

I felt like I was doing better, but I was ever mindful that chemo is cumulative, and I was acutely aware that life had been a bit more challenging lately. I was closing in on finishing my first twelve-week round of chemo, hopeful that there would be good news before I started the AC round in September.

I'd been warned by Dr. Z and my oncologists that the side effects of Adriamycin and Cytoxan were much tougher than those of Taxol and carboplatin. Adriamycin is infamously known as "Red Death" or the "Red Devil." With a name like that, it cannot be fun. Curious as to why everyone referred to the

Adriamycin as "Red Death" or "Red Devil," I was finally told it was because of its deep red color and its high level of toxicity. It is designed to kill as many cancer cells as possible in your body. It's like sending in the whole damn U.S. Army along with the Navy and the Air Force. But once it's inside your body, it doesn't discriminate, so it kills off a lot of the other cells at the same time, including the good cells—the white and red blood cells that you need.

Interestingly, my oncologists and surgeon felt that my cancer might be completely gone by the time we got to the surgery in November. If that were the case, I would still have to follow the surgery with radiation. I wanted to remain as optimistic as possible. The good news just kept pointing me in that direction, because I got the results from my gene testing: Everything came back negative for BRCA1 and BRCA 2 and the entire panel. This meant I did not carry any genetic predisposition to developing breast cancer known to date. If I'd tested positive, we would have to be concerned about the possibility of ovarian cancer as well, and it would impact my daughters' risk factors for breast and/or ovarian cancer. Everyone was able to breathe a little easier when the results came back negative.

As bad as my blood numbers were, I was feeling surprisingly well. I was staying fit and strong by working out with Beth every morning and doing whatever I could think of to prepare for the arrival of more than two hundred Camp Reveille ladies later that week. I began to ponder whether it would be a good idea to skip my next treatment—which coincided with the first day of Reveille—and give my body a week to recoup a little, especially since my next dose came with carboplatin, which had the tendency to wipe me out.

My biggest worry with running Camp Reveille was having enough stamina to do a lot of the fitness classes with the women, like I had in the past. My oncologist was more concerned about my exposure to the germs of two hundred people. With my blood counts so low, getting sick could spell real problems for me.

There would be lots of photo ops, hugs, handshakes, and other compromising situations to think about and plan for, to make sure I didn't inadvertently expose myself to someone else's illness or germs. It's tough, because a person's natural instinct is to reach out and take your hand or hug you, especially in an intimate setting, like camp. I'd been dealing with it all summer, and it hadn't been easy, since I'm the kind of person who is quick to give a hug in return.

The women who come to Reveille are often surprised that I am so accessible and actively involved in all of the activities. I try to be everywhere so that my campers feel like they've worked out or roasted a marshmallow with me. I didn't want to appear aloof or distant this year, but I needed to self-protect.

Overhearing me discussing this concern, one of the fathers attending the father/son weekend, who works as an executive with Sony, told me that a few of his celebrity clients buy flesh-colored surgical gloves to protect them from everyone else's germs when they're out in public. He said I could find them at Costco.

Who knew?

As the clock ticked down, I had a lot of bunks to get ready and cabins to prepare prior to the ladies' arrival. All of my wonderful fitness trainers and mind/body facilitators were back, and I had some terrific speakers lined up. I was planning a healthy cooking class with Dr. Z. I also planned a "Live Younger Longer" panel discussion that would include Dr. Z, my fitness guru Beth Bielat, and Dr. Cheryl Woodson, a geriatrician I'd met in Chicago while speaking at a Washington Post Live forum on caregiving earlier in the year. She had a wealth of information about what made older people sick and unhealthy, and she pulled no punches when she spoke; I loved that. For a more holistic view, I had Dr. David Coppola, a therapist who specialized in emotional healing and lifestyle coaching. Each provided a completely different view on how to make healthy lifestyle choices and stay younger longer.

I also invited Jene Luciani, who wrote *The Bra Book* and could always lead a fun, wild, crazy session on "Do You Really Know Your Bra Size?" Most women have no clue! She has the ladies write what they think their bra size is on a sticky note and paste it on their chest. Almost all of the women discovered they were off in size and wearing an improper bra. And finally, I had security expert Tracey Vega coming to do a session on personal security, from the information you give on your social media that makes you vulnerable to thieves, to where to park at the mall, to how to break out of a hold if you are attacked. There was a lot to do, but as always, I had quite a posse to assist me.

While every year of Camp Reveille is special, there was incredible meaning behind our weekend this year. I think all of the women were wondering how I would be doing, but I was determined to be there and to be as strong and as healthy as possible. As their camp leader, I wanted to create an outstanding and unforgettable experience for everyone.

By the time all of the fathers and sons left after their weekend of fun, the amazing Takajo maintenance and counseling staff went into overdrive to turn this boys' camp into a ladies' retreat. They had three days to go from summer camp to bed-and-breakfast.

Believe me, that was no easy task.

I'm not sure even David Blaine could pull off such a feat!

Thankfully, it didn't take a magician. We've got the most incredible housekeeping staff; they swiftly go from bunk to bunk and bleach them out, getting rid of any bathroom smells that groups of seven- to fifteen-year-old boys may have left behind. Our team then dusts and oils all of the wood shelves and paints the cabin floors, leaving everything looking brand-new. Next, they paint all of the shower houses a fresh coat of white and put up brand-new curtains at each shower stall. I put wicker three-drawer dressers into the shower houses and fill them with every ladies' amenity you can think of, from tampons to Q-tips.

Flowers are placed on the cabin nightstands, along with reading lamps. Plush two-ply toilet paper is placed in each bathroom, with a cute little basket of extra toilet paper and lots of potpourri and odor-removing spray. We also supply bug spray, sunscreen, and ear-plugs in the Reveille welcome bags. Each bed gets a special memory-foam pad, which completely trans-forms the sleep experience for the ladies, along with our exclusive Camp Reveille linens and comfy throws at the end of each bed to give the bunks a homey look and feel.

Take my word on this: The boys of Camp Takajo wouldn't know what to do if they saw how the ladies of Camp Reveille lived in those bunks for the weekend!

Camp started in a day, and I was definitely not feeling myself. I was usually running all over, doing the work of ten people, but I simply couldn't now, and that worried me.

By this time, all four of our younger children had returned from camp and were spending the week with us. On their first morning back, they were scheduled to take a tennis lesson with the camp instructor while I worked on the final details of who was bunking where and with whom. It was important to me that the women who shared bunks had lots of meaningful things in common.

While the kids were taking their lesson, it occurred to me that I needed to show them that I was doing okay. I wanted them to see me as normal. I didn't want them to think I couldn't be there, playing tennis with them. Besides, I had lost a significant amount of weight and wanted to show them how good I looked. I threw on my cutest outfit and headed up to the courts.

On the way over, I noticed I was unusually out of breath, though I thought nothing of it. But shortly after we started to volley, I lunged forward for a shot and tripped. I thought I had the physical stamina to do what I wanted to do, but I couldn't react as fast as I thought I could. Instead, I fell on my left hand with my full body weight, which really hurt my left wrist and palm, leaving me bruised for weeks. (Chemo causes the body to bruise much easier than it usually does.) I also scraped my knee so badly, it was bleeding. I pushed myself up and tried not to make a big deal out of it, saying, "My bad, my bad." However, I had really hurt myself.

In retrospect, I was far more emotionally bruised and battered than physically hurt. I had put myself in a terrible position, especially with two hundred women due to arrive in a couple of days. I didn't want anyone to know just how bad I felt, but the truth is, I felt awful.

I decided to walk back to the house before losing it in front of the kids. I could feel the lump in my throat, and I wasn't sure I'd be able to push it back. About five minutes into my walk, I let it go. This was the first real emotional breakdown I'd had since hearing the words "You have cancer."

Oh yeah, I'd cried after reading that email about the woman dying from breast cancer and leaving her young children behind, but I hadn't broken down like this.

It was a good one.

I just wanted to be normal.

I wanted to believe everything was okay, and it wasn't.

I sobbed uncontrollably, something I rarely do.

But I did it alone, in the absence of others, so no one would know how broken I felt.

Why?

Because this was the first day I felt cancer kicking me in the butt.

And I didn't like how that felt.

At all.

When I got back to the house, I called Jeff to let him know what had happened.

When he walked in the door, he found me still crying. He took one look at me, smiled, and said, "I thought you promised to take it easy."

Jeff then took me by the hand, led me to our bathroom, ran a hot bubble bath, and said to take as much time as I needed. "You have to acknowledge your cancer and your fatigue. If you are going to function at the level you want to be at for Reveille, you need to protect yourself and not do anything else stupid—not that playing tennis with the kids was stupid, but you have to conserve your energy for the next few days, or you're not going to make it."

He was so understanding and right.

With all of my angst about Reveille and my lack of energy, I was relieved that Jeff was taking me for my next chemo. It was the first time he'd accompanied me since I began seeing Dr. Weisberg. They hadn't met yet, and I was eager to have him hear her thoughts on whether I should take a pass on my next treatment or if she had some other plan. I was certain she would agree that with everything I had on my plate, taking the week off would be the right course of action.

Once again, they took my blood and we waited. Dr. Weisberg looked at my numbers and said, "If any other ordinary person was sitting here, I would probably rush to give them a blood transfusion. You are not any other ordinary person. I'm going to give you your full dose."

While my white and red counts were still very low, they were not low enough to panic or put off my treatment. She advised me to eat right to make up for my low counts. The good news was that my platelets were up—in fact, they'd doubled, going from 99 to 188, which meant I was doing all of the right things with my eating and supplement plan. With that good news, I got my treatment.

In the early afternoon, the women began arriving. I lay low until my first appearance at six P.M., during the "Meet and Greet" cocktail party. I made my way through the crowd, chatting with as many new people as I could while trying not to appear distant, since I was avoiding handshakes and hugs. Jeff stayed right by my side and made sure I wasn't compromised in any way. I'm usually so stoic, strong, and self-confident, but I will admit to feeling a bit vulnerable. Just feeling his strong presence next to me was so comforting. It was unusual for him to be with me at one of my Reveille events; after all, this was a ladies' weekend. I'd always felt that since I'd asked all of them to leave their husbands behind to come to camp, then it wouldn't be right to show up with mine—even if he did own the place.

Everyone could feel the excitement in the dining hall during the opening dinner as the women waited to hear what was in store for them over the next few days. After dinner, the group walked through the camp, along a tiki-torch-lit path, through a wooded area, into a huge open-air arena called the Indian Council Ring. As the women entered the arena and took their seats around the roaring campfire, they were greeted by tribal drums and chanting from a local musical group called Inanna. The mesmerizing troupe is made up of five Maine women who have performed at Camp Reveille for the past eight years. It is a favorite camp tradition for them to beat their drums and chant as the guests file in and take their seats around the campfire.

I had written my usual welcome speech; however, when I stood and tried to speak, I suddenly decided to go from the heart. I looked around the circle of women before me and shared with them how this incredibly beautiful setting had played such an integral part in my breast cancer journey so far and how it had inspired me to awaken each day and walk the grounds and stay fit—which had become quite important in my recovery. I also shared how this unexpected challenge had forced me to step back, exhale, and reevaluate my life, my career, and the amazing newfound relationships I had forged with so many Americans going through the same thing.

I told them how I was receiving hundreds of emails, tweets, and Facebook messages every day wishing me well, sending love and prayers, sharing tips and advice and their own journeys, and how it was all so meaningful and impactful and healing for me. I was aware that many of these women had been following my progress, and I felt I should let them in on how I was feeling, what was fueling me, and what had been inspiring me along the way. After all, the weekend was about being inspired and staying motivated. Not only was I bonding with them to enhance their experience at Reveille, but I also felt I had an opportunity to serve as a real role model on that unique evening.

Although it isn't a breast cancer event, we always had many breast cancer survivors among the group, and I knew that we had several women going through the journey that year. One woman in particular was having a very tough time with her chemo and had contemplated not coming. However, the six women coming with her were encouraging and supporting her. Even so, she was worried that she might be up sick in the middle of the night and would wake the others in the bunk. I told her not to worry, that I would put her in private housing with a bathroom so she'd have privacy, and then she could meet her bunkmates at breakfast and spend the rest of the day with them. I put an extra bed in her bunk so if she and a friend wanted to actually rest at rest hour, they could.

Her bunkmates who invited her to camp came to me as they were getting ready to leave. Through her tears, she told me how much it had meant for them to have this time together. The two of us hugged and had a good cry. This bond I was feeling, this sisterhood, this moment when two women were holding on to hope that we would have good outcomes, this was what it was all about. This was what the *People* magazine cover would be about.

After a lot of back-and-forth over the course of the summer, I had decided to allow *People* to send reporter Emily Strohm and photographer Greta Rybus to follow me around camp for the September cover story. I had been somewhat reluctant: First, I didn't want their presence or the photo shoot to interfere with the camp experience. And second, I was worried whether I would have the stamina to be a part of the activities without pushing myself so hard that I created a bigger problem for my health.

My initial idea was to have the *People* crew shoot me on the climbing wall, but after joining a thirty-minute cardio class and a camp walk, I felt fatigue setting in. There would not be any shots on the climbing wall. But I paced myself carefully, and the photographer caught nice shots of me walking amid the great pines on the grounds of Camp Takajo with groups of campers, doing the early-morning workout down at the beach, and running around like a crazy kid during our "whacked-up relay." This is such a unique opportunity for the ladies to act like they're twelve again, challenging each other at an egg-andspoon race, or a three-legged race around the bases of a ball field, or any other idiotic fun challenge we can come up with for them. This is the highlight of camp for many of the ladies, because for an hour or two, they are able to forget that they are grown-ups with tremendous pressure, stress, and lots of responsibilities on their shoulders. They simply get to be playful kids again, and oh, what a relief that is!

Camp Reveille turned out to be a huge success. When all was said and done, I heard from so many of the women that the long weekend would have a lasting impact on them. One woman said to me with tears in her eyes, "You don't realize it, but you saved my life with this weekend."

Another wrote and said she had reflected many times on our weekend together; she had been a part of the group who came to support their friend going through a personal journey like mine. She wrote about the reading we all participated in around the campfire, and how she was allowed to share her thoughts on faith with the group. She now strives to make the women a part of her daily life, and they remind their friend to *rest her body, remain positive,* and *continue to focus on her goals through courage and strength*: inspiring words they'd heard from the ideals I shared with the group on the night I spoke from my heart. She called the ideals I'd shared that night at camp the "Foundation of My Life."

Wow.

They should be the foundation of all of our lives.

It was sometimes difficult for me not to tear up, because I could feel their collective compassion and hope for me. At some point over the weekend, nearly every woman said to me, "You will make it, you are strong, and I will be praying for you."

I hoped I was as strong as everyone kept saying I was.

My dad, a young doctor at his desk in the 1950s.

We were a flying family, as seen here in our Christmas card from when I was a little girl.

MERRY CHRISTMAS

JOAN   JEFF   GLADYCE   ERLE

BLUNDEN JR

1959   BLUNDEN SR

My mom was always big on "expanding our horizons" and we took many trips as a family.

Me with my dad and his nurses in 1955.

My first day as host of *Good Morning America*. The press all wanted to meet Baby Jamie so she was brought to me on the set.

My early days at *Good Morning America* with cohost David Hartman.

Sammy Davis Jr. has a cuddle with baby Jamie while on the set of *Good Morning America*.

My three older girls often joined me on set. *L to R:* Sarah, Jamie, and Lindsay.

A special moment with my dear friend Charlie Gibson on the set of *Good Morning America*.

Lifelong friends. It's always good when I can get back together with my pal Charlie Gibson.

My rock. My husband, Jeff, and I on our wedding day.

Jeff and I with our first set of twins, Max and Kate, in Florida.

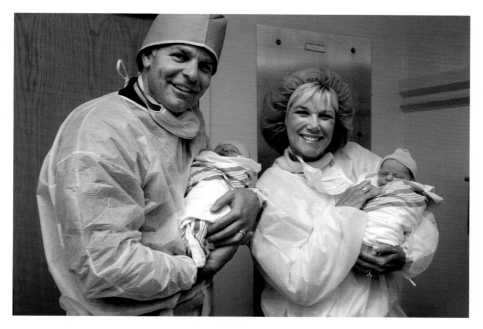

Jeff and I holding our newborns Kim and Jack, a second set of twins!

Right after receiving my cancer diagnosis I hosted a joint birthday party for my younger twins and a family baby shower for my daughter Lindsay. No one at the party knew yet about my breast cancer.

Kate and Max graduate elementary school (I was secretly in the midst of a flurry of oncologist appointments).

The whole family. *L to R, front row:* Kate, Jack, Kim. *Middle row:* Lindsay, Sarah, Max, Jamie. *Back row:* Joan, Jeff.

A family pile-on at our house!

Making my breast cancer diagnosis public with Robin Roberts on the set of *Good Morning America* on June 24, 2014.

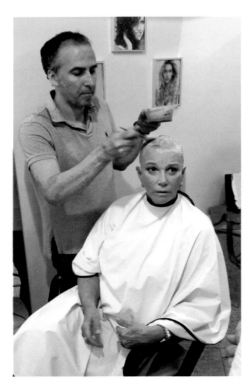

"I know I just have a little stubble and it's going to fall out anyway, but can't it be platinum?!" Emir Pehilj, my hair/makeup artist and close friend was like an angel to me throughout this past year.

My "G.I. Joan" moment as I get my head shaved.

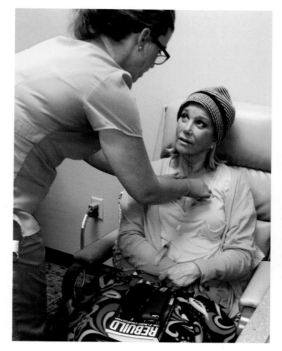

Oh c'mon, it couldn't have hurt THAT bad!

My nurse Jenny starting my chemo infusion.

Last day of chemo treatment.

Lasers lining up for radiation.

Hoda and I sharing a moment before sitting down to discuss my diagnosis for the *Today* show.

On set for our special #pinkpower series on *Today*.

My whole family surprised me on set for the last day of #pinkpower to show their support.

"Here's the story . . . of a hairless lady . . ."

My workout buddies at Camp Takajo and my fitness trainer, Beth Bielat, who kept me going every day.

In my wig and dressed in pink at my Camp Reveille.

Camp Reveille opening-night campfire offers inspiration and camaraderie.

With my older daughters at Lindsay's baby shower, which I threw in the middle of one of the toughest parts of my chemo treatment.

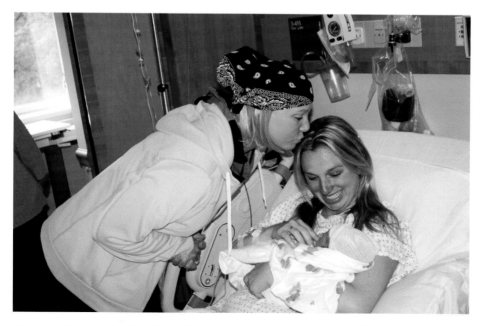

My baby, Lindsay, holding her baby daughter, Parker, after giving birth to her as I was finishing chemotherapy.

A private moment without my wig at a recent photo shoot with good friend and esteemed photographer Andrew Eccles.

Leading the Breakaway from Cancer "Survivor Mile" with fellow advocate Patrick Dempsey.

Out on the trail, speaking at one of many cancer events this past year.

A uniquely special moment as my husband, Jeff, presents me with the Spirit of Life Award for the City of Hope.

With Nancy Brinker, founder of Susan G. Komen, about to receive the Komen Impact Award, presented to me by fellow survivor Amy Robach.

Mistress of Ceremonies at the Honoring the Promise Gala at the Kennedy Center in Washington, DC, the night before going to Capitol Hill for the first time as an advocate for women's health.

My dad in the light suit, unveiling a "Fight Cancer" billboard for the American Cancer Society in 1960.

Leading a sea of pink at a cancer event in my own mission to fight cancer. At every event I go to, I feel like my dad is right there by my side.

Chapter 17

A Surprise Visit from Charlie Gibson

Don't miss your life . . .

VALERIE HARPER

*Actress, diagnosed with lung cancer in 2009
and brain cancer in 2012*

The end of August brings a peaceful presence that I look forward to every season. Instead of the sound of campers squealing with delight or cheering each other on, all I hear is the sound of the lake lapping on the shore. However, there was a lot to do before closing things up for the season and heading back to Greenwich. I needed to schedule my remaining chemo sessions in New York for this first round of treatment.

I was a little nervous about going back; I had gotten so comfortable with all of the wonderful and caring nurses in Maine.

Most people go to one facility to do their treatments, and that's it. Of course, by now you know my life is anything but simple. The time had come to say good-bye to Maine and get back to the city and my other team of professionals, where, despite my angst, I knew I would be in very good hands.

One thing I was certain of: I would genuinely miss the team of professionals at the New England Cancer Specialists. They made me feel so relaxed that I never once dreaded going to chemo.

That says a lot.

Change is hard, especially in the middle of something like chemotherapy. But as they say, the only thing that stays the same in life is change.

You can count on it.

After the ladies of Camp Reveille had left, Jeff and I decided to stay for a couple of quiet days. This was the first time the younger kids had been around me for any extended period of time since I'd gotten sick. I guess it's fair to say that it was also the first time I'd been around them.

I noticed the way they looked at me every time I came out of my bedroom wearing a different bandana or wig. Their little eyes would go to my head, and I'd know it represented the cancer that I was dealing with—and that they were dealing with.

I didn't want them to be scared or to feel awkward in their own home or about their mom being sick.

It felt so uncomfortable whenever Max looked at me with his worried, inquisitive little eyes. It made me very self-conscious. I knew cancer had changed the way I looked, and for their sake and their comfort, I needed to show up with some alternative hair or hat so I didn't worry them. I tried to make light of it, sometimes twirling in a circle, asking, "How do you like this new hairdo? Isn't it a hot, happening style?"

I was using humor to cover up the uncertainty I felt once the kids came home from camp.

Much to my delight, a week before the start of Camp Reveille, I received an unexpected email from my former cohost on *GMA*, Charlie Gibson, saying that he and his wife, Arlene, would be coming to this area to visit friends who lived about a mile down Long Lake from us. I'd been telling him about Camp Takajo and how special this place was ever since I met Jeff in 1996 and Charlie and I were both still working on *Good Morning America*. God, it seems like yesterday. It's hard to believe it's been eighteen years since I left *GMA*.

I remember when Charlie first began. David Hartman had been the longtime host, but the network was looking to make a change and was trying out several potential male hosts. Charlie had been a correspondent with ABC News for many years and frequently appeared on *Good Morning America*, primarily reporting on congressional news. He sat in as a substitute host for David Hartman on a couple of occasions, and I thought he would make a perfect new partner. Charlie was a cute guy, with a boyishness that made you instantly like him. He wasn't a pretty-boy handsome that would make it hard to take him seriously, telling you the news at seven A.M. He was just right. Charlie was also really smart, so I knew he would easily handle the challenges of two hours of live television every morning, when you never know what you could be talking about next. As a bonus, Charlie and I hit it off—there was wonderful chemistry between us. We genuinely liked each other, and that was so important. I believe it also showed.

When Charlie was officially named the new cohost of *GMA*, he came in my office and closed the door behind him. He wanted me to know that he was aware I had supported his being brought on the program and how much he appreciated it.

Humble, straightforward, and real. That's the kind of man he is.

He said, "Let's always be fifty-fifty partners. Let's not fight over stories. Let's show America how a male/female team can compatibly work together."

That would be a big change from how all of the morning shows were set up, where the male host always got top billing over the female cohost. I got along great with David, and he was always very nice to me. David had come to *Good Morning America* from Hollywood, where he had starred in several successful prime-time shows. It was a very different time when *Good Morning America* went on the air in 1975. David was cast as the star of the program, and since the show was originally run by the ABC Entertainment division, it was not so unusual to contractually make him the only star. There were many people in the news department at ABC and other networks that took issue with casting an actor in a news role. But to David's credit, not only did he stick it out, he persevered, and the show took over the number one spot in the ratings. I first began working with him in 1977, and while I was told to stay in his shadow, I will say that I learned a lot from David.

What Charlie was proposing was something very different. He was suggesting equal footing between the two of us, which created a very exciting, fun, and promising future together.

The week when Charlie started on the show, we were airing live from Fisher Island in South Florida.

Charlie's dad came along to enjoy his son's first day on the network morning show and share in his newfound success. Charlie was very close to his dad, a relationship I came to admire very much over the years.

That first morning Charlie and I cohosted *Good Morning America,* we opened the program on a beautiful South Florida beach. Even at seven in the morning, the hot tropical sun was already beating down on us. I had chosen an outfit appropriate for the beach shot, but conservative Charlie was wearing a navy blue suit and tie.

He looked *hot*—as in sweltering in the sun.

At one point I turned to him and said, "Let me loosen your tie, Charlie."

Which I did, much to his surprise. I don't think anyone at ABC News would have done that.

"You're not on Capitol Hill anymore," I said with a big smile.

For a moment he may have questioned his decision to leave his role as an ABC Washington correspondent to host this wake-up program, but once we got going, I don't believe he ever looked back.

After that first day, Charlie and I took his father out to dinner to celebrate. I took the opportunity to ask his dad what Charlie was like as a kid. His father spoke glowingly about Charlie, how he'd always loved history and learning about our government and how he was reading the *Congressional Record* when he was only eight years old.

Why didn't this surprise me?

That dinner certainly showed me that the network had picked the right guy to host our show.

I've always respected Charlie, and that night was the beginning of a wonderful relationship that blossomed and bonded over the course of our many years working together. We had the best working relationship anyone could ever ask for, on- and off-air. Charlie and I were like brother and sister. One thing is for sure, we always had each other's back. We protected each other, and like siblings, we also teased each other.

Oh yes, there was some good-hearted and spirited bantering over the years. If I noticed that Charlie had accidentally put on one brown and one navy sock getting dressed in the dark at four A.M. (it happens more often than you can imagine!), you can bet I brought it up on the show and had the cameraman get a good close-up look at his mismatched socks.

Charlie, our weather reporter, Spencer Christian, and I all joked, teased, and pranked one another whenever we got the chance. It was all heartfelt fun. Charlie used to joke to people that he and I were kind of like an old married couple because "we could finish each other's sentences, we just didn't have sex." Come to think of it, Charlie—*both* made us an old married couple!

There was such a bond between all of us—a trust that was unspoken. Our viewers must have known that while we treated our positions and responsibilities on the program with great respect, we were also having a really good time every morning. That was why we stayed number one for seventeen years.

Whenever Charlie and I get together these days, we reminisce about our years and the privilege we were given to wake up America. They are fond memories for both of us.

It was a breathtakingly beautiful day in Maine when Charlie, Arlene, and their friends from down the lake piled on a golf cart with us for a tour around Camp Takajo. This was a slice of my life that Charlie had heard about for years but never seen. I think it gave him a lot of joy to see how at peace I was, even with my current situation.

We had a wonderful catch-up visit. When it was time for a giant goodbye bear hug with my wonderful friend, Charlie held me a little tighter and a little longer than usual. He didn't need to say a word. I understood exactly what it meant.

"I'll be fine," I said in a soft whisper.

Chapter 18
Getting Back to Real Life

I kept my diagnosis under the radar, even from the cast and crew of The Sopranos, *because well-meaning people would have driven me crazy asking, "How are you feeling?" I would have wanted to say, "I am scared, I don't feel so good, and my hair is falling out." But I bucked up, put on my Carmela fingernails, and was ready to work.*

EDIE FALCO

Actress, diagnosed with breast cancer in 2003

Shortly after Charlie and Arlene left, Jeff and I locked up our summer paradise and drove south to Connecticut. With the kids in the car, we were headed back to what I usually referred to as "our real life."

I didn't even know what that meant anymore.

For most of the drive, I couldn't help thinking about how much and how fast my life had changed in just a couple of months. When we got to Greenwich, it would change even more.

And that terrified me.

All summer long, I had been living in my very neat and secluded world on the lake where I was able to throw away everything in my pantry that wasn't on my clean-eating plan and fill it and my refrigerator with only healthy choices. I was virtually living on my own. I could selfishly wake up when I wanted to, nap, come and go on my schedule, and make choices *for* me that were *about* me because my children were safely ensconced at sleepaway camp.

Now I would be back in Greenwich, in a house filled with young kids, lots of chaos, and loads of temptations. I had been eating healthy meals all summer, and now I would be tempted by pasta, chicken wings, and cheeseburgers. And did I mention pizza?

The protected little world where I had been able to stay healthy easily was about to be challenged in a big—okay, make that *huge*—way.

During the summer, I also had Beth Bielat knocking on my door every morning. I totally understood what a luxury she was, and now I would need to make sure I stayed as active and physically fit so I didn't lose the benefits of our hard work together. I knew how critical my active life and clean eating had been to the success of the aggressive chemo regimen so far. I didn't want to do anything that would compromise those results so close to the finish line. I needed to constantly remind myself that my summer was an anomaly. I was damn lucky to have had that time. Had my diagnosis happened at any other time of the year, the circumstances of my treatment may have been very different, because I wouldn't have been able to duplicate the intensity, focus, and single-mindedness with which I was able to approach my treatment in Maine; my "real" life wouldn't have given me the same opportunity. In Maine, I didn't have to figure out meals, schedules, lessons, work, obligations, and everything else a busy working mom deals with on a regular basis, which, even under ordinary circumstances, can be enough to take me down. I'd given myself a concentrated effort for three months, regardless of how protective my husband was and would continue to be, but you can't stop life from happening around you.

I had it good.

Really good.

Most women going through breast cancer treatments have to put on their wig, don a smile, and show up at work every day regardless of how they feel. Many of them don't tell their employers and colleagues that they have cancer or are undergoing treatment because they're worried it might affect their job security or chances for advancement. So they suffer silently, needlessly, without the support of their friends and coworkers, out of pure fear. The more I thought about that, the more upset I got. Of course, the more grateful I was, too, for having had the advantage of a very casual lifestyle for the first three months of my treatment. I'd had a playground to help make that time more tolerable. "Blessed" doesn't begin to state how fortunate I felt to have had that time and space.

But now, as the wheels of our car rolled me closer and closer to "real life," I, too, would revert to business clothes; I, too, would go back to work; I, too, would be wearing a wig and a smile to work every day. And like so many others fighting this battle, I would be going to school functions.

Max and Kate were starting a new school, which meant a new-student orientation for parents and kids. All I wanted was for everything to be as normal as possible for them.

Life was so incredibly simple up in Maine.

But real life is not always simple.

Chapter 19
A New Baby Brings New Joy

I look at everybody differently. I look at every child differently. I look at every flower differently. I am grateful for every day. It's like before and after . . . once you've had cancer, you just appreciate everything.

SUZANNE SOMERS

Actress, author, diagnosed with breast cancer in 2000

As summer was coming to a close, so was my first round of chemo treatments. I was scheduled to have my eleventh treatment and my first back in New York City just before Labor Day weekend. Sleeplessness had proven to be one of the more annoying side effects of chemo, which had made me perpetually tired. My need for sleep hadn't been helped by my efforts to stay properly hydrated, which translated to getting up several times at night to pee.

I was very nervous about seeing Dr. Oratz for the first time in nearly two and a half months. I was worried that my blood counts might be so low that she and her team would take a dim view of giving me my treatment. Dr. Weisberg had closely monitored the fitness plan Beth had mapped out for me and the food and supplement regimen Dr. Z had me on. She'd been confident that even if my blood count was low, when she gave me the treatment, I would build myself right back up. I needed to bring Dr. Oratz up to speed so that she, too, would understand how hard we had worked to build me back up every week after my treatment. I didn't want to lose any momentum or confidence in my progress.

But as I suspected, after my blood was drawn, it indicated that my counts were down. Dr. Oratz spoke with Dr. Weisberg about the situation and decided I could have the eleventh treatment, but she would not give me the twelfth and final treatment of my chemo. She explained that my blood counts had been too low for too long; my bone marrow shouldn't be pushed any further. She felt that my body had had enough for the time being. The best thing for me to do was give myself a break.

She cautioned me to take the next month and build myself back up before starting the AC round of chemo, which was scheduled to begin on September 22. As I had heard so many times before, Dr. Oratz warned me that I could expect the AC round to be far more challenging. She wanted to go over my schedule to make sure I was being realistic about what I was committing to for work, especially speaking engagements. She was genuinely concerned about my plans to travel around the country on crowded commercial flights while my blood counts were so low. Knowing that I was the typical type A personality (you know, the person who believes she's invincible and can do anything), Dr. Oratz was vehemently shaking her head as I laid out my overbooked schedule for the fall. She quickly reminded me that everyone reacts differently to chemotherapy but that going forward with AC, I would most likely be wiped out. I needed to understand that my immune system would be way too vulnerable to viruses. So while I could keep some of my prior commitments, she was insistent that I cancel several of the speeches.

This wasn't the news I had been hoping for. Throughout my entire career, I've never canceled a speech. I took it as a professional blow because I had made commitments to people that I couldn't keep. I despised the idea of letting people down, going back on my word, and being unable to be there for an event.

There had to be a way to salvage the situation. I didn't want to leave anyone disappointed by pulling out at the last minute. That went against everything I stood for as a professional speaker. Then I remembered that Leeza Gibbons and a few other colleagues had written me thoughtful notes when I first announced my cancer diagnosis. Each had offered to stand in for me if I was unable to make an appearance. Well, I sure hoped they'd meant it, because I needed someone, and fast. I reached out to Leeza first, and she was an absolute doll, as always. She immediately agreed to fill in. While the event planners were initially disappointed by my inability to be there, they were very nice and understanding of my unusual circumstances.

The following day, I got one of the shots they give after chemo infusions to bolster white blood cell production. Once it goes to your bone marrow, it's like a drill sergeant calling up the troops to go to battle and mobilize as many white blood cells out of the bone marrow as possible to counteract all of the white blood cells that the chemo is killing off. While all of this is going on in your body, you feel awful. You can have pain in the larger bones in your body—mainly your lower back and hips—where the biggest battles are being waged. The pain is due to the fact that these are the sites where adults have the highest content of bone marrow and therefore the highest mobilization of the white blood cells. Sure enough, right after I took my first shot, I suddenly came down with what felt like a terrible case of the flu. I had been out shopping for some items for the house when I felt such terrible pain in my hip that I could barely make the drive home. I had never felt such sharp pain, and at the time, I wasn't sure what caused it. Later, I understood it was the shot doing its job. I'm so thankful that I live at a time when we have medications that allow us to undergo toxic chemo treatments, though I wish they didn't make us feel so nauseated, achy, and headachy. I began feeling worse with every passing hour. I was so happy to get under the covers, close my eyes, and take a nap. Napping was something that was new to me but had become a welcome pleasure, especially at moments like this. I didn't want to go into Labor Day weekend feeling so lousy. As I dozed off, I thought, *Maybe I'm just having sympathy pains for Lindsay;* she had called earlier, saying she wasn't feeling great, either. The mother-daughter connection is so strong between us. *Of course, that must be it,* I thought before falling asleep.

Later that afternoon I was rousted back to consciousness by another call from Lindsay, who thought the reason she wasn't feeling well was because she had gone into labor.

Wait, really?

Labor?

But she's two weeks early!

I told Lindsay it was most likely false labor pains, or Braxton Hicks contractions.

Lindsay said, "I know it's early, but I'm telling you, these are real labor pains—real contractions."

She said she had already called Evan, who was on the golf course. Since she was planning to deliver at a hospital five minutes from our home in Connecticut, they would leave New York City when he got home.

Lindsay and Evan arrived at our house around eight o'clock that night. By that time, her contractions were about six minutes apart. Evan called their doctor, who thought they should come right over to the hospital. Lo and behold, an hour later, Lindsay had been admitted to the hospital!

This was no dress rehearsal.

I could hardly believe it!

Despite how lousy I had been feeling all day, Jeff and I started preparing to go to the hospital. Everyone knew it could be a long night, but there was no way I wasn't going to be there for my daughter or the birth of my first grandchild. I kept quiet about how horrible I felt and muddled through by putting one foot in front of the other and wearing a smile to mask my pain.

I don't think I told Jeff how miserable I felt, for fear he wouldn't have let me leave the house that night. Deep down, I knew he would want me to rest and take care of myself; he also wanted everything to be perfect for Lindsay. Neither of us wanted Lindsay worrying about me at a time like this. We wanted all of her focus and attention to be on the glorious occasion at hand— the birth of her first child.

When we arrived at the hospital, we were able to see Lindsay and Evan and spend several hours with them, until one-thirty in the morning. The doctors finally told us we should go home and get some rest. They thought Lindsay wasn't likely to deliver for several more hours.

Sleep sounded pretty good to me, but I was torn about leaving Lindsay's side. Evan's parents had arrived from Baltimore, and we all agreed it was the smartest thing to do. Evan promised to call the moment there was any news.

The house phone rang at five-thirty A.M. When your phone rings at that ungodly hour, it's for one of two reasons—good news or bad news. Joyfully, it was

Evan calling to say that Lindsay had just delivered the baby! We should all come back to the hospital!

We jumped up and dressed as quickly as possible and headed straight back to Greenwich Hospital. I was so excited by the call, and admittedly half asleep, that I forgot to ask the baby's name, which they had been secretive about for months.

When we arrived, Evan was in the waiting room. He took us to the maternity ward to meet *his daughter . . . our granddaughter . . .* a beautiful seven-pound, three-ounce baby girl named Parker Leigh.

When I caught my initial glimpse of this gorgeous little girl, my first thought was *She is SO tiny, SO cute.*

When the nurse took off her small knit hospital hat, we were surprised to see that she had a full head of dark hair. That was a bit of a shock, because Lindsay and Evan are both dirty blondes.

"Where did that come from?" we all joked.

And is it bad that I secretly envied that gorgeous head of hair?

Throughout the day, friends and family were called with the joyous news. One by one they came to visit and meet little Parker Leigh. I smiled and did my best to put up a brave front but sat somewhat lifeless in a chair off in the corner of Lindsay's room. I said very little because I felt so awful that I could hardly maintain an okay exterior. If I could have, I would have lain down on the floor and closed my eyes. In retrospect, I should have gone home, but I didn't want to miss a moment of the celebration. I finally left the hospital that afternoon to get the second shot to bolster my white blood cell count. Then I went home to get some rest.

Well, Labor Day sure took on a whole new meaning for us that weekend. However, once the excitement of welcoming Parker Leigh into the world had passed, the rest of the weekend was a complete write-off for me. I was completely out of it, with severe stomach pain, nausea, headaches, and aching bones—all side effects of the medications I was on. Thankfully, Lindsay and Evan had so many visitors, I was hoping they didn't notice my absence. At least I hoped they didn't.

I struggled with terrible bouts of guilt for feeling so bad during these seminal moments of Lindsay's life. Her pregnancy had coincided with my diagnosis. It should have been the happiest time of her life, upbeat and full of celebration and frivolity, not downtrodden and full of doctor visits and chemo treatments with her mom. I knew she didn't mind, and there was nothing anyone could do about the timing—which sucked—but I would have much preferred going to ultrasounds with her, listening to little Parker's heartbeat, over sitting in a chair getting chemo any day of the week.

Lindsay has since told me that she knew how bad I was feeling the day Parker was born, but she had Evan, her family, and so many other people there for her, too. She didn't want me to worry any more than I wanted her to worry. She said what mattered to her most was how excited I was about the new baby. And despite how lousy I felt, she was so happy that I was there. Hearing this from Lindsay filled my heart with joy and contentment. It makes me so proud to know that as her mom, I've done a pretty good job. And now that she's a new mom, I know (or at least hope) she will pass on that wisdom and knowledge to baby Parker. Honestly, that's the greatest gift any parent could ask for—that and a new grandchild.

Chapter 20
A Bold, Bald Move

I feel it's important to make a mark somewhere.

EVELYN LAUDER

American businesswoman and philanthropist, who established the Breast Cancer Research Foundation and formalized the Pink Ribbon for Breast Cancer Awareness, diagnosed with breast cancer in 1989

As the blissful celebration of Parker Leigh's birth was winding down, the reality of real life was setting in. The twins were heading back to school, with Kate and Max starting middle school. I was going back to work after a peaceful summer in Maine, and Kate Coyne, an executive editor at *People* magazine, was calling to discuss my upcoming photo shoot for the cover. She wanted to know if I would consider going on the cover of the magazine . . . BALD!

What?

Seriously?

They were sending a photographer to my home in early September to shoot for a late-September cover, just before Breast Cancer Awareness Month in October. Kate Coyne was pressing hard for me to do the cover bald, saying she felt it would be iconic and could make a big difference for women with breast cancer everywhere. They wanted to show the approximately three million women undergoing chemotherapy that their worst fear—hair loss—was nothing to be ashamed of. She made a point of saying, "If Joan Lunden can conquer this considerable hurdle and look amazing while doing it, then so can any woman." She was sure that if I said yes, it would have a powerful and positive impact all over the world.

WOW!

This was a lot to take in, especially coming off the emotional high of the weekend I'd just had, coupled with feeling so physically bad. Although I was starting to come out of my chemo-induced funk, I was on the fence about the request from *People*.

The health advocate in me was incredibly excited by the impact the cover could have, creating positive discussions about breast cancer. However, the personal and vain side of me felt totally scared, insecure, and vulnerable.

Would going bald on the cover of *People* really give a voice to tens of thousands of women struggling with cancer?

Oh, man—I hoped so.

I just wanted to fight this fight against cancer and survive; I didn't want to be embroiled in any controversy over my battle to live. I had already endured my first round of chemo and its crummy side effects for eleven weeks, having to look in the mirror every day and see myself bald. How would I feel with my bald head on the cover of a magazine for the whole world to see? Very few people had ever seen me bald. Allowing the entire world to see me bald was so intimate and vulnerable, making it one of the scariest things I could think of at that moment—scarier than going through my next round of chemo.

For real!

How would I do this?

Should I do this?

Was I brave enough to do this?

Gee whiz, I wasn't sure.

It became another huge decision as I made my way down this uncharted path.

I discussed it openly and frankly with my grown daughters, who thought it was *mostly* a good move for me—though they were clear that they would stand by whatever decision I made. Looking back, I think there was some hesitation, especially from Lindsay, who was being hyper-protective of me. She told me more than once that it was okay to nix the idea, because she could see how much angst I was feeling over it. The reality was, I didn't *have* to do it.

So then why was I vacillating?

I couldn't say yes, but I couldn't say no, either.

That wasn't my usual MO. Things are usually very black or white for me.

Living in the gray was taking some getting used to.

My husband knows me well, so he completely understood my dilemma of wanting to help others while feeling embarrassed to let everyone see me bald. He felt that doing the cover bald had incredible potential, but like my daughters, he worried about the stress the decision was creating for me. If I was feeling this much pressure now, what would it be like when the magazine hit the stands?

That was definitely food for thought!

I had a lot of discussions about the cover over the next several days. It seemed like everyone had an opinion. No single point of view felt right or wrong. I received lots of correspondence from friends, family, and even my oncology nurses.

Dear Joan,

I have been thinking about our conversation all evening regarding your *People* mag cover shoot. You asked what I would do. My true honest answer is I would wear my wig. You are beyond brave for sharing your story with the world. No one will think less of you for wearing what makes you comfortable on the cover. I think you are worried about protecting your family, and there is no need to add stress to an already trying time. You should be proud of all the many lives you are touching! Just wanted to share my honest opinion since you asked. I hope you have a restful weekend. No option is right or wrong, whatever you choose.

I was really touched that my nurse had been thinking about our discussion long after I had left the treatment center, and was worried that I was stressing over the decision. So I wrote her right back:

Hi,

I really appreciate your email. Haven't made the decision yet, I guess because of my 20-year connection with the American people as host of *Good Morning America* I still feel a responsibility to them. Funny how that happens, it has been over 18 years since I left, but it was such a strong connection and I hear from them ALL the time, and they are so loving and so well wishing, and just so appreciative of all the years in their homes. So just because I'm not on *GMA,* it's not like I walk away and never hear from them and forget them, because we live in a world of social media and the Internet, so you stay connected to them and continue the relationship.

However, I do agree that it is a stressful decision and that I don't need the stress, especially right now. And frankly, what HAS TO BE THE MOST IMPORTANT CONCERN IN MY DECISION is how my children, 9-year-old twins and 11-year-old twins, will feel about it. But again, I really appreciate your concern and your wonderful connection with me on this journey. It has made this journey so much less difficult. Thank you so very much!

Joan

I remember once saying to the oncology nurse that I always felt like a bald girl isn't fully dressed without some funky knit cap on her head. We laughed when I said it, but then I shared how weird I felt being bald in front of my husband. While I knew in my heart it (probably) didn't bother him a bit—he's so incredibly loving and supportive—let's face it, he married the bold, blond TV host Joan Lunden. I couldn't help myself: I genuinely felt strange and uncomfortable whenever he caught a glimpse of the bald me. At first, when I took off my wig at night, I slipped a little cotton knit hat on my head. I told him it was because I got so cold when I didn't cover my head—and since I'd started chemo, I was always freezing, so that was true. But quite honestly, it was because I was initially so embarrassed to let him see me bald. He could say it didn't matter, that it was no big deal, but I had two eyes and a mirror, and I could see that it was a big deal. Sometimes I heard that Rod Stewart song "Do Ya Think I'm Sexy?" playing in my head, and then a resounding NO! being shouted back.

Okay, it did make me laugh.

Thank goodness I still had my sense of humor.

And that was the whole point of doing the *People* cover without a wig. By putting my bald image out there for everyone to see, I would be sending a positive message: "Yes, I'm bald, and yes, it looks very different, but I'm smiling because I know I'm doing the right thing."

Still, I was grappling with how Jeff would really feel when everyone, everywhere, saw his wife with no hair. If I chose to do this, would that be fair to him?

And what about the kids?

I needed to know how they felt about it, too.

So one night we all gathered on my bed, and I threw it out for discussion. We talked about the pros and cons and how they would feel to see their mom bald on the cover of a magazine.

Max was the first to raise his hand. His arm shot right up immediately as I posed the question; he said, "You should do it, Mom. It will help a lot of people."

And with that, all of the kids agreed.

Ultimately, the decision rested squarely on my shoulders—just like my bald head. I had to decide whether I was courageous enough to pose bald on the cover of *People,* or whether I would weenie out.

But wait.

Appearing bald on the cover of *People* could become the ultimate representation of strength and power over cancer.

How could I ignore that opportunity?

Well, that sure sounded good and even made me feel a lot better about doing it.

Ugh!

Stress is not good when you are battling cancer, especially when you're dealing with the emotional and physical side effects of chemotherapy.

And then I had an epiphany. I knew exactly what to do. I would hold off on making any decision until the day of the shoot. I would simply let my heart and instincts guide me. They hadn't let me down yet.

Chapter 21
No Turning Back Now

I do have a fear of failure. I cannot fail. Not when the stakes are so high, with so many lives in the balance.

NANCY BRINKER

CEO of the Susan G. Komen Foundation

There was a lot of excitement in our house as Kate and Max got ready to start their first day of middle school. I set the alarm for six A.M. so I could be awake and ready to take them to their new school. On the car ride there, Jeff and I did our best to make them feel comfortable and at ease. We couldn't tell whether they were excited or nervous about going to a new school with kids they didn't know. I sometimes think it's the parents who are more anxious in these situations.

As for me?

I was excited about what lay ahead for these two great kids—they adapt easily and make friends wherever they go. But I was also contemplating the potential fallout a bald *People* cover would present for them in their new school. They were at an age where kids say cruel things. I didn't want them to endure any type of ridicule if I chose to do the cover without my wig. I didn't want to knowingly create any embarrassment or distress for Kate and Max or any of my children. Above all, I didn't want anyone to ask, "Is your mom going to die?"

Oh, this was creating so much angst—and the clock was ticking down, because it just so happened that today was the day of reckoning.

After we dropped the kids off and said our goodbyes, Jeff and I drove off like it was any other day.

But it wasn't.

"Are you ready?" he asked.

That was the ten-thousand-pound elephant in the car, wasn't it?

Kate Coyne and her crew from *People* would be at the house in a few hours, wanting to know my decision. I was trying to conjure up the courage to take the shot and make the statement. The only hurdle still holding me back was how vulnerable and uncomfortable my family might feel. Even so, I took a deep breath and told Jeff I was leaning toward doing the cover bald.

Without missing a beat, my incredibly supportive husband was quick to assure me that he thought that was a brave decision and he was proud of me.

Whoa.

"Are you sure that you won't feel embarrassed about everyone else seeing me bald?" I asked.

Jeff reached across the front seat, took my hand, and said, "You are beautiful with or without hair."

By the time we got home, I thought I had made up my mind to go forward, but once again, I was having second thoughts. There was still something pulling at me. Kate Coyne and Brenna Britton, the deputy photo editor, asked to talk to me before I totally backed out. We went into my bedroom and into the closet where I kept my wigs. As we stood there talking, Kate shared a very personal story about her grandmother, who had been diagnosed with breast cancer but chose not to go through chemotherapy because she was afraid of losing her hair. As a result, she died. If she had seen a cover like the one I was contemplating, perhaps her outcome would have been different.

That did it for me.

I don't know if Kate Coyne even had a sick grand-mother, but she sure is good at her job.

"I can do this, Kate! I'm ALL in," I said.

There was no turning back now.

The *People* shoot was very different from any other I had done. This one had purpose—intentionality. I knew where I was headed in the outcome. In the end, there would be a message in that photo. If it's true that a picture speaks a thousand words, then the images that photographer Ruven Afanador captured that day could have filled a novel.

When I shared my decision with Kate, I wanted to be sure that she understood I was doing the cover bald for specific reasons. It was important to me to give the thousands of women out there going through this battle a voice, to show them that just because you lose your hair during chemo, you can still be strong, vibrant, and normal.

I needed them to know that as soon as the chemo is over, your hair will grow back.

That it's not that big of a deal.

Husbands, partners, family members, and friends need to be as supportive as possible, because we do feel weird and kind of embarrassed as we are going through it, but we can and will come out of it—as survivors!

And that is what matters!

If one woman who was neglecting treatment because of a fear of losing her hair and possibly her breast(s) gets this message and seeks treatment and it saves her life, then my being seen bald for a few moments—okay, days; okay, maybe even weeks or a few months, since the picture would probably live on virally for a while—WILL HAVE BEEN WORTH IT.

There was a buzz and energy in the air as we got started. Ruven began the photo shoot with a wig, then with a beautiful Hermès scarf wrapped around my head: a nice accessory to the beautiful Helmut Lang ivory sweater they had me wearing.

When it was time to pull off the scarf and take the photos without my wig, I asked that they clear the set of everyone other than the photographer and those absolutely essential for obtaining the shot. My daughter Sarah stayed to keep a close eye on the setup and to make sure she liked what was being shot. I think Ruven was pretty psyched when he looked through his lens. We were all aware this was a most unusual circumstance.

As I stood just a foot or two in front of the camera lens, extremely conscious of the camera's close proximity, I thought about my dad and how proud he would be. He would say, "Smile effusively so that you connect with everyone and make them feel comfortable with what you are doing. That's how you will get the chance to make a difference in this world, my baby doll."

That's what he always called me . . . baby doll.

I can't say that I felt much like a baby doll in that moment. I felt more like a fierce warrior, a brave and mighty warrior taking a stance, driving home a point, and showing everyone that my uncomfortable decision might send a message that would be like a wave across a nation, hopefully changing the way we see women dealing with breast cancer and maybe changing the way we deal with the ladies in our lives going through this challenging journey. When I looked in the lens, I needed to wear that bold attitude to pull it off.

Something told me I had when I found Ruven looking at his computer screen. About an hour after we finished up, we were shown four different options of the "bald" photos that could be used for the cover of the magazine.

I will never be able to find the words to thank Kate Coyne for helping me find the courage to allow this *People* cover to be shot. Her support and vision for the cover was spot-on. Although I wasn't positive what the response would be, she was. From the start, Kate totally understood the reaction her readers would have. I had to believe she knew something I didn't. It was a giant leap of faith—but one I had to risk taking, in honor of all of the warriors fighting breast cancer.

Chapter 22
The Unexpected Curveballs of Cancer

Too many women are so afraid of breast cancer that they endanger their lives. These fears of being "less" of a woman are very real, and it is very important to talk about the emotional side effects honestly. They must come out in the open.

BETTY FORD

First lady, diagnosed with breast cancer in 1974

I had my first appointment with Dr. Barbara Ward, my cancer surgeon, since leaving for Maine at the beginning of the summer. During my physical exam, she surprisingly said, "I don't really feel your tumor at all. I clearly felt it the last time I examined you on June twelfth, but I don't seem to feel it now. I want you to have an ultrasound and see where things stand."

Wow!

Really?

Does that mean it might be gone?

If that was the case, then what would be my next steps?

Okay, I would be lying if I told you that this didn't get my hopes up. I mean, come on! My cancer surgeon said she couldn't find my tumor!

Was it possible that it really had gone away?

Did we really kill it off over the summer with the chemo?

Would I even need more chemo, and if I did, how many rounds?

If the tumor were gone or almost gone, I could skip more toxic chemo and go straight to surgery.

Okay, wait. I didn't want to get ahead of myself and possibly set myself up for disappointment.

Breathe, Joan, breathe!

I would have to patiently wait for the results of the ultrasound, because that would dictate everything going forward.

I wasn't sure what to expect when I went for my ultrasound. I went in trying to treat it like any other appointment. As always, I took off everything from the waist up, put on the gown, and lay down on the very same table where they had discovered my tumor a couple of months earlier. Having any type of procedure where people are looking for tumors is nerve-wracking, but there was a glimmer of hope, and unlike last time, I had a sense that there might be some good news when the test was over.

The technician came in, opened up my gown, instructed me to put my right arm up over my head, squirted that cold squishy liquid onto my right breast, and started to make her way around my chest. She seemed to be going back to one area of my right breast that they hadn't looked at before—an area not even close to the location of my tumor.

I immediately freaked out and became invested in my fear that she had found another tumor. This was exactly what had happened during the last ultrasound: The tech kept going back to the same spot again and again.

All I could think while I was lying there, vulnerable and exposed, was: *Please, God, don't let her find another tumor. We're supposed to be killing cancer cells, not growing them! How could this be?*

Talk about paying dues on a debt not yet incurred!

Just lie there, Joan, and try not to think, I reminded myself as I tried to drown out the horrible thoughts racing through my mind.

Then the tech started poking around in the general area of the tumor. She appeared to get frustrated. She finally looked at me and said, "I'm going to call in the doctor now," and left the room.

I was left lying there, almost paralyzed by the thought that she might have found something else. A few moments later, Jeff came in to be with me. I didn't tell him what I was thinking or that she may have found another tumor (because all of this was in my mind).

Dr. Calamari came into the room with the tech, picked up the ultrasound wand, and started looking at the images on the screen.

The tech looked at the doctor and said, "See what I mean? I can't find the tumor!"

The two kept searching and finally saw the clips that the radiologist had put in when she did the initial biopsy, but little else.

The doctor looked at me and said, "It seems as though the tumor is almost completely gone. It appears to have shrunk about eighty percent in size. This is terrific! You have had really amazing success!" It was what we had all hoped for, but it was almost too good to be true. And you know what they say: "If it's too good to be true, then it probably is!"

Therefore, when I heard this news, I felt somewhat conflicted. Was it the greatest news I'd ever heard, or did it leave us wondering about the future? That's the thing about cancer. You want to believe it's gone, but it's a tricky foe.

The following evening, Dr. Oratz called to talk about my ultrasound results and our next steps. She was really impressed that the tumor seemed to have shrunk about 80 to 90 percent. She explained there was a possibility that I wouldn't need any more chemo treatments.

NO MORE CHEMO?

Wait!

Did she just say NO MORE CHEMO?

Yes, she did.

She said I might be able to go straight to surgery for a lumpectomy, which would be incredibly minimal, with a short recovery, since there was essentially very little tumor left.

"We did this together," she said.

And we did.

My response to the treatment spoke volumes for using carboplatin in addition to Taxol, but the way I had been staying fit and on my clean-eating plan had also greatly contributed to my success.

I hung up the phone and told Jeff this amazing news: NO MORE CHEMO!!!

We immediately called my daughters, and he called his parents, and we shared the good news.

This changed so much!

Unfortunately, my elation would be short-lived. While Dr. Oratz was optimistic about my tumor, she was still concerned that my hemoglobin was so low and had been for several weeks. She said I might need to consider getting a blood transfusion so I could get back on track to being "my old self."

Back to my old self . . . boy, I liked the way that sounded. I'd forgotten what that even felt like!

My life felt like it was in such flux. I'd never been the kind of person who did well with instability. I was used to having every detail of my life in place, sometimes for as far out as the entire upcoming year.

Perhaps it's the Virgo in me, and my need to have order in my life, but I like things to be super-organized. Every entry on my calendar is color-coded: business appointments, travel, hair/makeup, press interviews, kids, doctor appointments, etc. I've been doing this for as long as I can remember—long before things were electronic.

But lately, my life had become so hard to predict, and that kind of uncertainty was a challenge for me. I had no way of knowing when or even if all of this would be over. I had no idea whether my cancer was gone—and if it were, would it come back someday? I was itching for my business and travel life to get back to normal.

To be completely candid, I really wanted all of my life to get back to normal.

Just when I thought that everyone had essentially written me off as "a cancer patient" who couldn't be booked for any work while fighting my disease, I got a rather unexpected email from Debbie Kosofsky, a senior producer with the *Today* show, asking if I would be interested in being a special contributor for an upcoming segment on breast cancer awareness.

What an unanticipated but super-delightful request.

They put two of their veteran producers, Yael Federbush and Brittany Schreiber, on the breast cancer series, which they were calling #PinkPower. I knew I would be in great hands.

But wait.

Would people question my doing this? It was the *Today* show.

My usual nature is not to upset the apple cart. Would I be doing that by accepting an invitation from the *Today* show? There was a rivalry between *GMA* and *Today* that had existed for as long as the two shows had been on the air.

Did accepting this role pose any dilemma for me after being associated with *GMA* for so many years?

I didn't want *GMA* to be upset with me; however, this was business. Was I never to work again on morning television because at one time I worked on *GMA*? I had an important message to deliver: I wanted to motivate women to be screened and possibly save lives, and here was the opportunity. Truth be told, I missed all those people with whom I spent every morning with for two decades. Maybe I couldn't see them, but I had breakfast with them every day, I had conversations with them, I shared all the joys in my life with them—they were like family to me. My connection to them had been cut off, and while I knew that many years had passed, it still felt almost like yesterday that I'd smiled and said "Good morning" to all of them. I imagine deep down, I wanted the chance to reconnect.

It had been a lot of years since I'd hosted *GMA*, yet I had faith that the American audience would be fine with me moving on. In fact, they might just cheer it! They might even say, "It's about time."

No, I didn't see any issues with making this appearance.

I had to go for it.

Now I really needed to get myself strong again. However, I had come to the same conclusion that Dr. Oratz was right. I was so fatigued that we did need to intervene. I couldn't even walk up a flight of stairs without stopping halfway to catch my breath. This was unlike anything I had ever felt. Perhaps for the first time in my life, I was feeling the effects of both my age and my disease. This was not normal—at least not *my* normal—and I didn't like the way it made me feel.

I will confess to being a bit skittish about the idea of having a blood transfusion and someone else's blood running through my veins.

Who was the person?

What was he or she like?

Was the procedure safe?

Even with my hesitations, if a transfusion would help get me back on the right path and help me find my energy again, I'd suck it up and do it. After everything I'd been through so far, what was the worst thing that could happen?

I called Dr. Ward's office and asked her to arrange for the transfusion to take place at my local hospital. She hadn't even heard the good news about my ultrasound. Naturally, she was elated, and her office gladly set up the blood transfusion for the next morning. First I would need to go to the blood lab in her medical building that day to find out my blood type. When the results from the blood test came back, we were surprised to discover that my hemoglobin had inched up a little, causing Dr. Ward to question whether we should go ahead with the transfusion or hold off.

However, Dr. Oratz knew what a rigorous schedule I had in the coming weeks, and she knew me well enough to understand that I really had to be dragging my ass to admit that I needed help, so she ordered the transfusion to go through.

I'd never had a blood transfusion, so admittedly, I was a little scared. When Sarah walked into my room at the hospital and saw the blood in the bag hanging above me, she looked at me and said, "Have you given any thought as to where that blood came from?" I must have shot her a "Thanks, I *didn't* need that" look, because she added, "Just close your eyes and hope it was from some tall, gorgeous, über-thin supermodel!"

I quickly countered with "Let's just hope she was a healthy supermodel!"

Just then an IV specialist came into my room, and prepared to go in through my chest port. She didn't have that awesome little spray bottle of local anesthetic that freezes the skin right before the needle jab, like my chemo nurses had. The IV specialist wanted the particulars on the make and model of my port so she knew what size needle to use.

I'm sorry, but that's just too much information to ask when you are speaking to a needle weenie and you're about to shove a needle into her chest!

She said she would just use the bigger one-inch needle. "Sorry, take a deep breath," and then she pushed it in.

OUCH!

Yes, it hurt—a lot.

But I used the good ol' Lamaze breathing method. I thought I had shelved that breathing technique many years ago. Yeah, turned out it's just like riding a bike, because it came right back to me when I needed it!

Sarah and I figured I would be at the hospital for an hour or two. Once they had me hooked up, the nurse told us it would take up to four hours per unit to receive the two units of blood I'd be getting. So that meant we would likely be there all day long and into the evening, for that matter.

Who knew?

Whenever I'd seen this type of thing on TV, it always looked like it happened so much faster!

Several different nurses came in and out throughout my transfusion and all told me that after receiving a couple units of blood, people usually feel much better. They said I might not be ready to dance out of there, but I ought to notice the difference right away.

I sure hoped they were right.

Chapter 23
One Lump or Two?

Time is shortening. But every day that I challenge this cancer and survive is a victory for me.

INGRID BERGMAN

Actress, diagnosed with breast cancer in 1973

With the curious findings of my ultrasound, Dr. Z was questioning why I was waiting to have my lumpectomy. Dr. Oratz and Dr. Ward agreed that it was a good idea to get me on the schedule for the week of September 23. Before surgery, I needed to have a complete blood count (CBC) and an EKG to make sure I was in good enough health to handle surgery.

"No problem, I'll have those done as soon as I get back from Washington!" I gleefully said.

You see, I had agreed to emcee the Honoring the Promise Gala for the Susan G. Komen Foundation for the Cure at the Kennedy Center on September 18. Nancy G. Brinker, the founder, had been incredibly supportive throughout my journey. It was my great honor to be a part of the annual black-tie gala, which brings together the heroes of the breast cancer movement for a celebratory evening of inspiration and world-class entertainment.

It had been a while since I'd gotten all dolled up. I was looking forward to it, since I spent most of my summer in workout clothes and sneakers. Thankfully, I had recently purchased a beautiful, elegant long black gown that fit me perfectly, except that it was a little long, so it needed to be hemmed. I wanted to find a hipper pair of heels to wear onstage, but much to my surprise, I had recently discovered that I had a problem wearing high heels; my feet just wouldn't endure heels and hadn't for a year or so.

Was it due to too many years in high heels or just another unexpected side effect of chemo, the gift that keeps on giving?

Either way, I needed to figure it out and fast, because there wasn't a pair of Nikes that I could reasonably pull off with that gown!

The other important event I was busy preparing for was my upcoming interview with Hoda Kotb for the *Today* show, scheduled to take place at my home. I had really come to admire Hoda and was very happy the show decided to send her to do the interview. It was fortuitous that she happened to be sitting in for Savannah Guthrie, who was out on maternity leave at the time.

Hoda had survived breast cancer and had gone public with her battle, so I knew I would feel comfortable speaking with her. However, I quickly downloaded her book, *Hoda: How I Survived War Zones, Bad Hair, Cancer, and Kathie Lee,* so that I would know the full story. I always research the person who will interview me; it has paid off time and again. I find Hoda really easy to talk to. She is genuine, sincere, down-to-earth, and so natural. These are all traits that make her a tremendous interviewer and someone who creates instant trust. However, there would be more than that at play during this interview. Not only did Hoda and I share the horror of a cancer diagnosis, we also shared the pain of divorce, and we had both lived our lives in the public eye. I was so at ease knowing that she would be the one coming to do the interview. I would be talking about human fear and the fight to survive, and having someone like Hoda there helped me feel safe in knowing I could reach deep down inside for that interview, which I did.

The following morning I was scheduled to fly to Washington, D.C., for the big gala. When we landed, a car swept me off to a hotel, where I had to quickly get changed for the big event.

Later that afternoon, I made my way through our nation's capital, winding past the Lincoln Memorial and the Washington Monument toward the famous Kennedy Center for rehearsals, and then for my walk up the Pink Carpet.

Driving through Washington brought back many memories from my *Good Morning America* days, from elegant dinner parties at the White House to live coverage of the presidential inaugurations. Inaugurations, in January, were always freezing cold. You just knew that as the morning wore on, your lips were going to be so cold that you could barely form words. I remember setting up our cameras during the wee hours in January 1985 for Ronald Reagan's inauguration, when the windchill was minus twenty degrees. Just before airtime, a decision was made to move the ceremony indoors. Good thing, since our cameras were having difficulty operating in the bone-chilling temps. I was stationed in the Capitol Dome, and that was where I met Charlie Gibson for the first time in person, when he was the capitol correspondent for ABC News. The inaugural atmosphere was chilly inside the Capitol, as Jimmy Carter was handing over the presidency to Ronald Reagan, and rumors were floating that Reagan had made a deal with Iran that would delay the release of the American hostages being held in Tehran until after the inauguration. In 1993, when Bill Clinton was taking over from George Bush, there was a two-day inaugural festival, with tents stretching from the Capitol to the Washington Monument and one million people attending the event. This was the lure of working on a program like *Good Morning America*, always being where history was being made.

Tonight's gala was taking place at another grand historical site, the Kennedy Center.Ordinarily, I would be there as just another celebrity attending as a "supporter" of the cause. But that night, I was one of the thousands of women for whom the event was working so hard to raise awareness and money to eradicate the disease we shared. Each step I took along the pink carpet that night represented my walk along the communal path. I was now proudly united in this fight against breast cancer.

Singer Julia Murney took center stage in the darkness to open the evening's festivities. As soon as the spotlight hit her, she belted out a rousing version of "Defying Gravity" from the hit Broadway show *Wicked*. Jeff and I had just taken our daughter Kate and a friend of hers to see that show in New York the weekend before, and I felt a surge of emotion build inside of me that mirrored the crescendo of the music. There were moments every day that I appreciated a little more because I realized just how precious, if not fragile, life had become. Perhaps life had always been that way and I was just figuring it out. I don't know, but I was really proud to be there that night—to willingly and boldly step out onto the stage and declare, "I have breast cancer, too," especially if it meant raising more money, more awareness, and giving more women hope.When I returned to Connecticut, I went in for my pre-op blood work and EKG the very next day, as promised. Dr. Ward said she was satisfied with my blood count, although she'd expected it to be higher after the transfusion. Unfortunately, she saw something of concern on my EKG. An EKG, or electrocardiogram, is a test that checks the electrical activity of your heart. I told her I had an EKG every year at Dr. Albert Knapp's office, and there had never been an issue in the past. Still concerned, she spoke with Dr. Knapp, who didn't seem to share the worry. However, Dr. Ward insisted on a second opinion before agreeing to let me have my lumpectomy.

Of course she did.

Why would anything along this journey be cut-and-dried?

She called another cardiologist, who agreed with her and felt I should have a stress echocardiogram before the surgery. I would need to do that test in the morning, before my lymphoscintigraphy test, which I was scheduled for later that afternoon. The lymphoscintigraphy test is when a radioactive substance is injected into the breast; it flows through the lymph ducts and can be taken up by lymph nodes. A scanner or probe is used to follow the movement of the radioactive substance on a computer screen.

Thankfully, I passed both tests with flying colors and was cleared for surgery the following week.

Chapter 24
The Best Birthday Present Ever!

Every year on my birthday I go out into the wilderness alone, to celebrate being alive.

LINDA ELLERBEE

Television journalist, diagnosed with breast cancer in 1991

September 19 is my birthday.

Has been my whole life.

But allow me to let you in on a little secret: I've never really loved birthdays.

For real.

Remind me why we celebrate them again?

I mean, after a certain age—say, ten—they suck!

Don't they just represent being another year older?

All right, all right, so my opinion about birthdays might have been slightly influenced by the battle I was waging against breast cancer (and perhaps that stupid Beatles song about turning sixty-four), but I wasn't much in the mood for celebrating this particular occasion.

Not this year, anyway.

Sure, I supposed I should appreciate and rejoice in the opportunity to be one year older.

And yes, I got a couple of really cool and thoughtful and funny gifts!

I got a T-shirt that said, "Cancer Touched My Boob So I'm Kicking Its Ass" across the front. I loved it!

I got a beautiful silk scarf that had words of strength and hope all over it. I'd definitely use that.

I got a tiny, easy-to-use head shaver called a Peanut, by Wahl. Oh, yeah. I'd use that, too, because I didn't want to use Jeff's razor on my head.

I got a couple of cute knit hats: my favorite gifts because it was getting cold out and we were headed into a brutal winter.

To be certain, the *best* birthday gift I got had come three weeks earlier, with the birth of my granddaughter, Parker Leigh. I was happy that mommy Lindsay and beautiful little baby Parker were doing great.

Okay, if I'm going to be reflective here, I would be remiss if I didn't appreciate that after a summer of aggressive chemo, I got a pretty awesome birthday present when my follow-up ultrasound showed nearly no tumor!

Happy birthday to me!

Happy birthday to me!

My upcoming lumpectomy would bring me much closer to the finish line.

So yes, this was one year I ought to consider celebrating.

And yet I didn't really have it in me.

A month before my birthday—well before I knew about the outcome of my chemo or the birth of Parker Leigh—I happened to mention to Jeff that I didn't want him to buy me anything. I just wanted to go to New York City for a night out with him.

I wanted a date night with my husband. "Dinner and a Broadway show?" I said.

I needed an upbeat, happy, irreverent show with great music. I'd heard from so many people that *Kinky Boots* was exactly that: a terrific, rousing performance. I thought it would be the perfect medicine for me.

Just before we left for the city, Sarah called us into Jeff's office, where she was working on his computer. It turned out that she had the one thing we were all waiting for . . .

THE picture.

The BALD picture.

The picture for *People* magazine.

Apparently, earlier that afternoon, Kate Coyne had emailed Sarah the following with the final image attached:

From: kate_coyne
Date: Friday, September 19, 2014 at 3:02 PM
To: Sarah Krauss
Subject: At last!

Hi there—

So, attached, please find the image we've selected for Joan's cover. It is, in fact, one of the ones we looked at when we were all together in her house, and we think it is really stunning. As you'll see, Joan looks healthy and radiant and happy—not sick, struggling or like a victim, and one of our staffers even pointed out that the shape she's making with her hands is almost like a heart. So fantastic! And that twinkle in her eyes . . . all there.

As I said earlier, some color correction is still ongoing, and the resolution and color you're likely to see on your computer screen likely won't do justice to what the final print of the image will look like. But this is essentially it: a beautiful, striking, empowering image that I just know is going to strike a chord with so, so many women. I mentioned to Joan when we met that I lost my grandmother because she refused to have chemo because she didn't want to lose her hair . . . I honestly think that if she'd had something like this to look at, she might have felt very, very differently.

Thank you again for all of your support and input and assistance; if there is anything else at all that you need, or if there's anything you want to discuss, don't hesitate to call me over the weekend on my cell.

My very best to you and Joan,

Kate Coyne

I looked at the picture on the computer screen with Jeff by my side. There were no words spoken for several seconds between us.

"I think it's going to be a good thing," I said.

Jeff loved what he saw, too. Without missing a beat, he said it was beautiful, striking, that I should be very pleased to look that beautiful with no hair.

As we stood there taking it all in, I think we were both surprised by the image that Ruven Afanador had captured.

Then Jeff turned to me and said I should be so proud of what I was doing.

I won't lie. I thought it was pretty awesome.

I came home later that night and found wonderful birthday wishes everywhere I looked. There were flowers and letters all over the house. They were there before we left for the city, but I was absorbing the meaning and the love and the reach of each and every one only in that moment.

When I popped open my laptop, I was heartened once again to find many more birthday wishes from people on Facebook and Twitter. The number of messages was mind-blowing. At first glance, I felt like I wanted to sit right down and try to get back to each and every one of them, but I soon realized that would be impossible. All I could do was sit back and appreciate them.

Maybe birthdays don't suck.

I began writing an email to several colleagues who had worked with me on different projects. Something inside me felt compelled to do it now, while it was on my mind and in my heart.

When I first went public with my diagnosis, Lindsay made sure to reach out to the CEOs of the companies I had long-standing relationships with, including Dr. Howard Murad, of Murad skin care, and Sean Kell, the CEO of A Place for Mom. I asked

Lindsay to assure them that I would still be fulfilling my work responsibilities throughout my cancer care, although I wouldn't recommend any photo or commercial shoots until I had a chance to grow back some of my hair. The reality of what I was going through, chemo and complete baldness, would be evident with the cover of *People* hitting newsstands. They would see the evidence of my disease in living color, and so would everyone else. I had represented my skin-care line, Resurgence, for Dr. Murad for close to a decade; I swear by it. I feel it has kept me looking as young and vibrant as possible, and I knew it was helping me get through my chemo. I constantly hear from women on social media that they have breakouts and terrible skin issues from the treatments, and I was experiencing none of that from mine. I was the face of his products, so I felt enormous pressure to always represent them well.

Surely the photo would speak for itself, wouldn't it?

Even so, I worried over how my important business partners would feel about seeing my bold bald shot. When the cancer treatment ended and my hair grew back, I still wanted to have those business relationships. Was I doing everything I could to preserve them?

And what about other people I did business with, such as agents and various speakers' bureaus? What about my friends and colleagues? I thought I should say something to them, too, before they walked through their supermarket and saw me smiling back at them with no hair.

I want to let you know ahead of time that I am going be on the cover of the *People* magazine coming out next week—with no hair. Yes, that wasn't a mistake; I was photographed for the cover with no hair. It was a tough decision for me, certainly not a comfortable one for me—to be seen bald—but I felt it was an important one in terms of empowering women across the country. I did it to make a statement: Chemo causes temporary hair loss—but it saves your life!

You all know that I have been a committed health advocate for years, and I knew in my heart that by sharing this challenging breast cancer journey, I could perhaps get more women to take charge of their own health and see that it's not the end of the world to lose your hair for a while as you get treated. The hair grows back and you can continue to live your life.

My hope is that it will encourage women to get checked, for early detection is key in getting treated and surviving.

Sometimes birthdays aren't just about getting. They are about giving, too. When I finished writing the note, I took a deep breath and hit send. I didn't think about it. I just pushed the button. Perhaps the greatest birthday gift this year wasn't what I got. Maybe, just maybe, it was what I was about to give.

Gosh, I sure did hope so.

Four days after my birthday, I was scheduled to have my lumpectomy. I started my morning with a quick visit to Dr. Calamari, the radiologist, who located the metal clips she had inserted at the tumor sites, and she attached two wires to them. She did this so Dr. Ward could use an MRI machine to find them inside the breast. I was very nervous, as it was not the most comfortable experience, but I was assured that it would allow for a potentially less invasive and smoother operation, so I grinned and bore it.

As I understood it, those wires were sticking out of my breast when I left the radiologist. I didn't have the stomach to look. I wanted to get right to Greenwich Hospital and check in for my lumpectomy. I was eager to get the procedure over with. If all went well, my tumor would be gone—as in over and out—when Dr. Ward was done with the surgery.

Dr. Ward explained that the procedure would take about an hour and a half and that I should be in recovery for around the same amount of time. If everything went as planned, I would be able to go home that day.

Since the tumor was reduced to almost nothing, she reassured me that the surgery should be minimally invasive. She would make an incision in my armpit, where she planned to remove the sentinel lymph node—even though the cancer hadn't spread to any of my nodes, it is usual procedure to take a couple of the ones closest to the tumor, just for good measure. The sentinel lymph node is the clearinghouse lymph node. Some women have one while others can have up to four or even five. All the lymph from the breast flows FIRST to the sentinel lymph node. If the sentinel lymph node is free of cancer, then there is little chance that the other (30–40) lymph nodes in the armpit harbor cancer cells. This procedure has made the axillary lymph node dissection essentially obsolete.

Next, she made another incision on my breast where she would go in to remove any remaining tumor, plus margins, and the clips that had been inserted during the biopsy.

I came out of the surgery a few minutes before two in the afternoon. Jeff and Sarah were there to meet me. All throughout my experience, I had never opened my eyes without a loved one staring back. I counted my blessings every day for that gift, love, and support. The more time I spent in hospitals, the more apparent it became that not everyone had that luxury. The thought of battling cancer alone truly broke my heart.

But I also worried about those who were navigating life alone, especially those dealing with cancer or some other chronic illness. Every time I sat for one of my chemo treatments, I became aware that there were countless people who were genuinely alone during their sessions. They weren't just sitting by themselves; they had no one to go through their illness with—no one to rely on for support or help. I heard from many of these brave women who were going through the journey on their own, and my heart went out to them every time I read one of their letters.

If you ever wanted to give the gift of your time, sit with someone going through chemo or radiation. Hold his or her hand, read aloud, perhaps take a home-cooked meal. I don't know that I would have understood before my illness how important these acts are, but that surely speaks to the change in my outlook today. The smallest kindness or compassion means so much to someone facing a chronic illness. Even if you can't give money to a cause, everyone has time to offer.

I was very groggy as I lay in recovery, but somehow I was finding the strength and peace of mind to be me.

What was the first thing on my mind?

Even though the holiday was a month away, I was trying to plan out my Halloween costume!

It was right around the time Dr. Ward walked by my room that I blurted out, "I want to be a surgeon for Halloween! In fact, I want to be Dr. Ward! I think she's fantastic."

I'm told I was really out of it and quite funny that day. I wonder what drugs they had me on!

Sarah, Jeff, and Dr. Ward sure seemed to have a good laugh.

I was really glad everyone had an intact sense of humor, because *People* was about to hit newsstands the very next day.

They were going to reveal the cover for the first time on the *Today* show.

Yikes!

Wasn't putting it out on every newsstand enough exposure?

They would be showcasing the cover on the morning talk show, too?

OMG!

What do I do?

Should I stay home all week?

Or hide for the next couple of weeks?

Will it be embarrassing to be seen out in public while the magazine is on newsstands?

I suppose the silver lining to this dilemma was that I didn't have to worry about being seen for a few days: I was home recovering from my surgery. I also had a family holiday dinner planned. The dinner had been scheduled long before my surgery, and I decided not to cancel it at the last minute. For the family's sake, the fewer changes I made to everyday normalcy, the better. Besides, I had thirty-two family members coming at six P.M., and somehow this felt *normal?*

I know. I know.

This was something that would make most people come undone, but I loved it!

I thrive on setting a mood, creating tables that will delight everyone, young and old, and making holidays memorable.

Oh boy. "Memorable" would be the understatement of the year, especially if I had a copy of me bald on the cover of *People*!

Would I have the chutzpah to show everyone?

What would they think?

They were my family . . . they had to be nice about it . . . at least to my face, right?

But what would they say about it in the car after they left?

I couldn't worry myself about it any further, because what was done, was done. They'd all find out sooner or later, anyway.

Chapter 25
The Big Bald Reveal

Now, if I go through it again, I think I would be a lot more open about it. I admire people who have been open like Melissa Etheridge and women I see walking around facing it without wigs and all that stuff. I think I would be more courageous next time.

KATHY BATES

Actress, diagnosed with ovarian cancer in 2003 and breast cancer in 2012

Jeff and I turned on Channel 4, our local NBC station, and watched the top of the eight A.M. hour of the *Today* show. Matt Lauer introduced Kate Coyne, who explained to the television audience that she was there to share a very special and unique cover of *People* that day—one unlike anything they'd ever run before.

Then she pulled back the cloth covering the large-board mock-up and revealed *the* cover.

I could actually hear gasps coming from the studio when my bald image appeared.

Gulp.

Kate said that I was sharing my breast cancer journey with the world and had chosen to do this brave cover to be a voice for thousands of other women fighting the battle against breast cancer. My goal in doing the cover without hair was to show everyone that you can still be strong and that your health is more important than your hair.

Moments later, I was patched in live with Matt and the other cohosts for a quick call to see how I was doing. I explained that I was recovering from my lumpectomy but feeling strong and hopeful. I sounded a bit groggy from surgery, but it must have created an impact, because this was when things started to blow up (in a good way) in social media. People from all over the country were reengaged in my journey. It was exciting and invigorating for me to feel their love, kindness, and support, just as I did when I made my announcement earlier in the summer. I was really looking forward to appearing on *Today* the following week and creating an even stronger bond with the audience.

After my lumpectomy, the question remained as to what to do next in my treatment: more chemo or not.

I spoke with a couple of experts and got a couple of different opinions about whether I should go on with chemo, since my smaller tumor was completely gone and the larger tumor had been reduced, seemingly by 70 to 80 percent. Could I consider myself cured?

I emailed Dr. Tracey Weisberg, my oncologist in Maine, for her opinion. She and I had really connected over the summer, and I was eager to hear her thoughts. When I received her answer via email, it struck a chord.

Dear Joan,

I think your situation presents some real issues for consideration.

First the facts:
1. Up to 17 to 20 percent of breast cancer is triple negative breast cancer.
2. Of triple negative breast cancer, the most virulent is the basal-like subtype.
3. Basal-like subtype usually (but not always) develops in young women, frequently of African-American descent.

4. Basal-like subtype breast cancers are frequently resistant to chemotherapy.
5. Triple negative (be they basal-like or not) may have unique capacity to spread through blood and avoid lymphatics.
6. Triple negative breast cancer—more specifically, the basal-like subtype—is frequently associated with germline BRCA1 mutations.

The specifics about your cancer:
1. It is triple negative, meaning no ER/PR/or Her2neu expression.
2. The initial size was estimated to be 1.8 to 2.3 centimeters.
3. With carboplatin/Taxol, you had a PR or partial response to neo-adjuvant therapy
4. The lymph node was negative for cancer. In triple negative, this may not be as reassuring an outcome as in ER-positive breast cancer due to the possibility of contamination of blood exclusive to lymph.
5. You tested negative for BRCA mutations.

About neoadjuvant chemotherapy:
1. Researchers are working hard to prove that having no identifiable tumor cells in the breast is a predictor of long-term survival. To date, this

has not been unequivocally proven in clinical trials.

2. From current clinical trials, it has not yet been *universally accepted* that Taxol and carboplatin added to AC should be the standard of care in *triple negative breast cancer.*

3. The BRCA-negative patients should be treated with dose-dense AC followed by Taxol (or ddAC-T).

At my center, we enrolled eight women on CALGB 40603. I was lucky. None of them were randomized to ddAC-T (control arm). All received carboplatin regimen plus AC with or without Avastin. To date, none have relapsed locally or distantly. We are two-plus years into follow-up.

So . . .

1. Based upon your result of having even microscopic residual disease, it would be very difficult for any oncologist to recommend omission of AC. Frankly, it would be difficult to recommend omission of AC even if you had had a complete response to therapy based upon current knowledge base, because remember, a complete response rate has not yet been 100 percent linked with assurance of cure and failure to relapse at a distant site.

2. There is little knowledge to date with regard to treating breast cancer exclusively with carbo-platin and Taxol.

So what to do . . .

1. The consequences of metastatic disease are devastating. Once cancer recurs in lung, liver, bone, or brain, there are drugs that can "hold it off" for a period of time. Those available drugs in triple negative breast cancer to date are exclusively chemotherapy and new drugs that are still investigational.

2. The toxicities of AC are manageable. They are not easy, but with good supportive care, women get through therapy with minimal "collateral damage." The biggest concern is potential side effects on heart, but frankly, with doses used in breast cancer, the true rate of heart toxicity is low if heart function is normal at the start of therapy. This is measured with a non-invasive test called an ECHOCARDIOGRAM. The other side effects are low white blood cell count (rem-edied with growth factor Neulasta) and nausea (managed with Aloxi, Emend, dexamethasone).

3. If you are a person who wants to do everything reasonable to beat this, I would do the AC. It

is a riskier position to omit this drug sequence based upon current evidence. In the future, science may make us smarter, but right now we are stuck.

4. If you choose to decline the AC, I would ask 1) do you consider yourself a lucky person, 2) can we make a strong case that you DO NOT have the most virulent form of triple negative breast cancer, 3) if the unthinkable happened (recurrence in lung, liver, brain) and you had declined the AC, would you forever "beat up" on yourself for not putting up with the further inconvenience of the eight-week period of the ddAC?

5. By the way, AC can be given one day every three weeks x 4 and may have the same benefits of ddAC.

I know that this is not exactly what you were hoping to hear, but I think your cancer team needs to do the best to cure you and make sure that you remain a citizen of the planet for absolutely as long as possible. Your team can support you through toxicities of future treatments.

You mean too much to too many people to do less than what seems to optimize outcomes in the current medical literature.

This discussion is always a bit clearer in person. Wish I was there in person to help you along. I am always available to you by email or phone!

Tracey

Wow!

DO I FEEL LIKE A REALLY LUCKY PERSON?

That question really stuck with me.

It was like I was suddenly playing cancer roulette.

Maybe there were no cancer cells left in my body, but then again, maybe there were.

Was I feeling lucky?

An image of Clint Eastwood in *Dirty Harry* popped into my head.

Was I willing to bet against the house?

Could I just go on with my life, walking around with my fingers in my ears, always hoping I was right?

No way.

I could never be that irresponsible.

I could never live my life that way.

I had to take the second round and consider it my insurance policy that I'd killed every possible cell that would have a chance to grow into another life-threatening tumor in the breasts or elsewhere.

Thanks to that email, I learned so much—especially that when it came to triple negative breast cancer, there was a tendency for the cancer to bypass the lymphatic system and go into blood vessels. So when everyone was reassured by the pathologist saying I had nothing in any of my lymph nodes, it may not have meant that much. I was more concerned that a cell or two might have flaked off and gone into a blood vessel and traveled elsewhere. And since I had triple negative breast cancer, a fast-growing aggressive cancer, it could get a foothold and grow quickly, so it was imperative to douse it with a strong chemo.

I was eternally grateful to Dr. Weisberg for her candor and caring ways.

Wow.

What a turnaround.

First I thought I needed the tough AC round, then the great news that I probably didn't, and now there I was, right back to the tough news that I must go through it after all.

Such is life—especially when facing death.

So much for growing back my eyebrows and eyelashes any time soon.

For real.

The night before my first appearance as a special correspondent for *Today,* as I was getting ready for bed, I washed my face. I had done that many times during my chemo treatment without incident. However, this time, when I looked up into the mirror, it appeared as though a part of my face had been . . . erased.

Gone.

My eyebrows and eyelashes had been washed away.

I woke up with eyebrows and eyelashes, and the night before I was set to appear on national television, they were washed down the drain.

Really?

The night before doing a breast cancer series on *Today?*

You've got to be kidding me.

C'mon!

I was completely shocked by the face staring back at me in the mirror.

It was the face of a sick woman.

A cancer patient.

I examined it almost as if it were someone else's face staring back at me.

It was SO weird.

I walked back into the bedroom to tell Jeff what had happened, trying to keep my composure, but he had already fallen asleep. I didn't want to wake him over something so trivial—I mean, it wasn't trivial to me, but given what I'd been through over the past few months, it could appear inconsequential to someone else.

I got into bed knowing and feeling that I really looked like a chemo patient now.

Oh dear.

What will people think when they see me tomorrow?

The *Today* show put me up at the Essex House hotel in New York City, where Sarah and I stayed for the entire week. Every morning Emir came to the room at five to do my makeup, sans wig. By seven, we were in the car sent by the show, heading to the Forty-eighth Street entrance to 30 Rock, the location of the *Today* show studio.

I was a little nervous, walking into Rockefeller Center that first morning. It was strange but not like I thought it would be. Everyone was so warm and welcoming that you would have thought I used to be the host of *this* morning show. There was absolutely a sense of community, of family.

Maybe it didn't matter what network you worked for. Perhaps we all belonged to that special fraternity who awakened in the dark and prepared to rouse the rest of America each day. Only a small group of us can say we've done that for a living. It's both an honor and a privilege, one I've been especially grateful to have had in my life.

People have asked me many times if it seemed strange being on the *Today* show set, but frankly, it felt like home. I know that may sound strange. But they made me feel like it was a "welcome back to morning television" event, so it felt natural as opposed to awkward. It was like getting back on a bike; I felt relaxed and at ease within minutes of the camera rolling. I give tremendous credit to Matt Lauer, Hoda Kotb, Natalie Morales, and everyone on team *Today* who went out of their way to make me feel so at home.

On my second morning there, show producers loaded the plaza outside Studio 1A with hundreds of women who had breast cancer and were currently going through chemotherapy. Many had come that morning proudly exposing their bald heads. I can't say if it was a sign of solidarity or a salute to my decision, but there sure were a lot of beautiful, bold, bald women that rainy and cold morning out on the plaza.

It was purely a celebration of awareness and the breast cancer community.

And oh, what a tight and loving community it is.

When I walked out onto the plaza before the segment began, I was able to take pictures and talk with a few about their experiences. THIS was why I was there. The connection I felt was deep and heartfelt. We had so much in common; there was a bond that tied us together in an indescribable manner.

That segment was great, and afterward, Sarah and I ventured across the 30 Rock concourse level for a hot hazelnut coffee, my favorite splurge. After stand-ing outside in the cold rain for a couple of hours, I was chilled to the bone and needed something to warm me up. Along the way, Sarah and I saw women wearing their #PinkPower T-shirts. I stopped to speak with all of them, but there was one who stuck out in my mind. She was about twenty-eight years old and had just given birth to the cutest little baby boy, sleeping in the stroller next to her. She was completely bald, proud, and so excited to be speaking with me. As we walked away, I couldn't stop thinking about her. She was a new mom, with so much life ahead of her. I was deeply impacted by our exchange.

She's so young, I thought many times throughout the day. I couldn't help but wonder how she'd found out about her cancer, given the guidelines to get your first mammogram at age forty. She was walking proof that this belief needs to be reconsidered. While it's still true that most breast cancers don't occur until later in life, more and more younger women are getting breast cancer. The question is why? Hormones, insecticides, pesticides, and chemicals in processed foods? Or is it environmental? Or could it be the chemicals absorbed through the everyday products we're exposed to inside the home, from those we use to clean, to the lotions we slather all over our skin? Even lipstick is on the list of potential dangers! It boggles the mind to think about the exposure and possibilities, and yet it is my belief that not enough is being done to alert younger women to look for the signs.

Later, after the segments on the plaza, Sarah and I went back to our hotel room and looked at the amazing response coming in on social media. What a new world that was for me. I was hearing from thousands of viewers all over the country! I was hearing not only their well wishes but also their concerns, their fears for their own health, their frustrations at not being allowed to have lifesaving tests, and their confusion about how to efficiently interact with the medical industry in order to properly advocate for their health and longevity. They had families and young kids who needed them—they couldn't afford to die and leave their families!

They needed help.

This was just one message I read that day demonstrating the confusion and the frustration and, in the end, the sadness and the positivity that can be found among women trying to get properly diagnosed and cared for.

My story starts back in February. My husband and I traveled to Australia to go to three Bruce Springsteen shows and drive up the Gold Coast. The night of our last Springsteen show in Brisbane, I got bitten a few times by what I thought was a mosquito on the outside edge of my hand. It itched but soon stopped. The next thing I know, it swelled up and was purple. It was similar to a blister, so I lanced it to try to dry it out. When I let my hand hang down, it hurt like heck and throbbed. I kept icing it, even as we started our journey home through Fiji. It had started drying out and healing by the time we got home on 3/4.

Over the next week I developed pain under my right arm; it was so painful I could barely lift my arm. I was also supposed to go to numerous functions, but declined because I felt jet-lagged. My husband talked me into scheduling an appointment with my doctor. I called on 3/10 and got in with Mohammed Ahmad Al-Hijji, MD, on 3/12. He was a resident, not my normal doctor. I was afraid it may be breast cancer. He and the director of internal medicine at Hopkins Wymann Park ordered a mammogram, ultrasound, and blood work. The mammogram was negative but the ultrasound showed two swollen lymph nodes under my right arm. I had the blood work done on Friday. So we were thinking it was related to the spider bite, not breast cancer. We scheduled a follow-up appointment for the afternoon of 3/19.

The morning of 3/19 I had an appointment with an orthopedist and was diagnosed with frozen shoulder on the same side. We met with Dr. Al-Hijji and the director that afternoon. They said my blood numbers were extremely low, white count, red count, platelets, etc. They were all extremely low. They said they wanted me to see a hematologist immediately, and to do that, they wanted me to check myself in to the emergency room and have them admit me. They said if I tried to schedule an appointment, it would take over a month and they didn't even want me to wait a week! They sent me on my way with a box full of face masks. I went home, packed an overnight bag, and did what they said.

The emergency room doctor said she didn't know why they were being alarmists but agreed to redo the blood work. They ended up admitting me just before midnight. On 3/20 I was diagnosed with leukemia. I began an aggressive course of chemo the next day. As it turned out, Dr. Al-Hijji was also doing his residency on the leukemia floor of the Weinberg Cancer Center of Johns Hopkins, it was nice to see a familiar face. I never cried and I didn't go into "shock." It wasn't until a week later that we found out I have a rare type called acute bilineage leukemia. Only 2 percent of leukemia diagnoses are this type.

Once they diagnosed me, they knew immediately that I would need a bone marrow transplant to survive. They knew early on my sister would be used only as an absolute last resort because she has MS; however, she ended up not being a match, so we had to go to the bone marrow registry.

As luck would have it they found me two perfect matches! I was scheduled to have a maxi transplant on 6/17.

I have also been type 1 diabetic since I was nineteen. Up until my first blood transfusion I never knew my blood type. My blood type quickly became my mantra. I actually had to explain it to a few people, but once they got it, they thought it was pretty cool.

Jennifer A.

When I opened up the attachment to the email, I found a photo of the blood from her transfusion. On the side of the bag, it read, "B (RH) Positive." In the midst of all the hysteria and fear, Jennifer found a focus in the "B Positive" on the bag of blood that was saving her life. And while her story isn't about breast cancer, it is indicative of two things I have learned along the way.

First, you have to be your own advocate on this journey. No one knows your body and what feels normal better than you do. If something doesn't feel right, say something. Tell your doctors. Don't hide it, because they can only base their determination on the information you are giving them.

Second, your mental attitude matters. An optimistic outlook, even when things look bleak, will help create a more positive outcome.

I was back on the *Today* show three more mornings that week, finishing out my special feature with a segment called "Ask Joan," during which viewers wrote in online and I spent most of the morning at a laptop in the Orange Room (that week it became the Pink Room), answering their questions during a live Facebook session. I wanted to answer as many viewers as possible during the first hour of the show, reading and typing at lightning speed. I loved the interaction.

In the third half hour of the show, I moved to the *Today* show desk to be with Matt Lauer. As the segment opened, Carson Daly was still in the Orange Room, along with Dr. Susan Drossman, a prominent New York radiologist. Carson read a couple of questions from Facebook, which I answered and then threw to the doctor for her expert medical opinion. After a few more questions, Matt turned to me and said, "We have one more question for you, Joan, and it comes from nine-year-old Kimberly from Greenwich, Connecticut. She wants to know why she can't have her ears pierced like her big sister."

Wait.

What?

I got it right away.

Matt might have been reading the question off a viewer question card, but he was definitely talking about my daughter Kim.

"Wait a minute, Matt, are we turning now to personal questions about my kids? That's got to be my daughter you're talking about!" I said.

Matt said, "Turn around and look behind you, Joan. Your entire family is here to support you this morning."

And there they were!

I never, not for one minute, expected the *Today* show to surprise me with my family!

I loved it!

I heard from so many people how wonderful it was to see my whole family together.

What a great ending to a great week!

Thanks, *Today* show!

Chapter 26
Wrestling with Cancer

Had I done the testing I needed to do, the treatment I would have gotten might not have been as aggressive. You don't save yourself anything by putting screening off. The breast cancer is either there or it isn't, whether you get screened or not. It does not change the reality. It only changes your options.

ELIZABETH EDWARDS

Estranged wife of presidential candidate John Edwards, diagnosed with breast cancer in 2006

There are some opportunities in life that just seem like too much fun to pass up, regardless of how lousy you might be feeling. Appearing on *WWE RAW*, on behalf of Susan G. Komen for a special event in honor of Breast Cancer Awareness Month, was definitely one of them.

Kathie Lee Gifford and Hoda Kotb were the scheduled guests of honor in the opening act that night at the Barclays Center in Brooklyn. I was supposed to come out in my personally bedazzled "Courage, Conquer, Cure" T-shirt, escorted by Triple H, to say hello and help introduce ringside breast cancer survivors. No doubt, this rocking, brawling night of wrestling was a dressed-down event, so I decided to wear a little knit cap over a hairpiece that was like a short "fall." I thought it looked casually rockin' cute and perfect for the occasion.

WWE called to say that Sarah and I would be picked up by a helicopter at the Westchester airport. While we tried to act cool, as if this happened to us all the time, we were quietly high-fiving each other when we heard about our slick ride. We were told we'd be sharing a ride with Kathie Lee and her assistant. That sure beat sitting in traffic, and it cut way down on the commute time. The helicopter made its way along the East River in the glare of the afternoon sun, landing on a helipad on the Lower East Side with the Statue of Liberty in the background. This was one of those rare New York moments that confirmed once again how lucky I am to live in proximity to one of the greatest cities on earth—that's how it must feel to be a rock star!

When we arrived at the Barclays Center, we were brought downstairs to an area in the arena where all of the wrestlers were gearing up for their show. We were escorted into a private dressing room where we waited for the evening to begin. At one point, everyone broke out into applause backstage. When we looked at the TV, we saw that The Rock was there for a special surprise appearance. It was a super-cool moment. While I was waiting to go on with all of the wrestlers dressed in what appeared to be their fancy underwear, Stephanie McMahon, the daughter of WWE founder Vince McMahon, came over to say hello. Vince McMahon got up from behind his monitor at the control center and came over to greet me, too, giving me a great big unexpected hug. The event was loads of fun and brought a little levity to the battle I was waging against this disease. Not just for me but for so many women facing the same prospects, perils, pain, and pressures.

The next day, I had to go from wrestling in the ring to once again wrestling with my cancer. I had a post-op appointment with Dr. Barbara Ward to check my incisions from the lumpectomy. They needed to be healed enough for her to give the go-ahead to the oncologist to start my AC round of chemo. While Dr. Ward examined me, she shared that the tumor had been difficult to find during the operation because it was so far back against my chest wall. She was concerned that I would be sore for some time, since she had to do "a lot of digging" during the procedure.

She removed the Steri-Strips to examine the incisions on my breast and under my right arm. She looked at my breasts and then looked straight at me and said, "The volume of your right breast is definitely smaller after the surgery. How do you feel about that?"

Interestingly, I had looked at them a couple of times, but the right breast had been so swollen from the surgery that I hadn't noticed a big difference. I guess I hadn't taken a really good look since the swelling had gone down.

Hey, it wasn't as bad as what so many other women have to deal with, from mastectomies to total reconstructive surgery.

I made light of it with the doctor, perhaps due to nervousness or maybe just denial. I didn't want to deal with that reality then. I told her I'd look at my breasts when I got home and I'd let her know how I felt after giving it some thought.

After the exam, I met Dr. Ward in her office, where Jeff was waiting for me. Dr. Ward said nothing about the smaller right boob. Instead, she focused on the success of the surgery, that I had clean margins and the tumor was out of my body. This was all great news.

Then she said she would need to see me for a follow-up in three months.

Just when I thought I was done . . . She hesitated and asked if I was scheduled to do a lot of appearances in the coming weeks. She had heard a lot of people around the hospital saying they were excited that I was going to be speaking at an upcoming educational conference at the hospital. The fact was, while I had canceled all of the October speeches that required me to travel on planes, I had left two local speeches on the docket. One was on Sunday, October 19, and the other was the speech at the hospital, which was scheduled for Saturday, November 8. Dr. Ward was clearly concerned that with just getting over my surgery and starting the next round of chemo, I might be pushing myself too much and, frankly, overdoing it.

Jeff, who is always very protective, agreed.

"Duly noted," I said to them both.

I mentioned that I was scheduled to do the *Today* show again the following week, on October 17, but after that, I was relatively free and clear of commitments. As a result of my schedule, we all decided it would be best not to start the next round of chemo as discussed but, rather, wait a week. Not only would that let me get those speeches and the *Today* show appearance out of the way while feeling better, it would give me a little extra time for my incisions to heal a bit more.

I could easily live with that decision, so I agreed.

I wanted to get started on the second round of chemo as soon as possible so I could get it over with faster. Thankfully, as long as I waited, Dr. Ward gave me the green light to begin, and with that, I was out of there and ready to begin the AC the following week.

While I was in the building, I met with my new prospective oncologist, Dr. Dick Hollister, to discuss doing my next round of chemo closer to home.

I'd already worked with two oncologists along my breast cancer journey, and I had been so happy with them. Now I would have to start all over again with another doctor, a new center, and new nurses. No one likes change. But I knew this was in my best interest. I had received such amazing treatment at the New England Cancer Specialists in Maine. Part of me wanted to get in a car and drive back up there for my next four treatments. Clearly, that was impractical and unreasonable, but the thought crossed my mind.

There was also a concern about whether he would go along with giving me the shorter, more intense regimen—dense-dose AC (ddAC)—in which you get infusions every two weeks instead of every three weeks, which was what I wanted. Although it's harder on you physically, this regimen would allow me to finish my chemo sooner.

I wanted to like this doctor. It would be incredible to find a chemo induction center in my hometown where I would be comfortable. Doing my chemo sessions close to home without the added pressure of having to go into New York City would make my life so much easier over the next few months.

When we met, I thought Dr. Hollister looked like the sweetest mad scientist you've ever met. He wore a long white lab coat, a bright bow tie, and had a white mustache to match. He was absolutely adorable, funny, and very smart. The first thing he wanted to know was whether I'd had a stress echocardiogram. This question telegraphed only one thing to me—that the AC round would be tough on my heart and might cause long-term damage. I'd just had this test prior to my lumpectomy, so I didn't need another.

Thankfully, I got a great feeling from Dr. Hollister. I could tell he had great compassion and cared deeply for his patients. Dr. Ward, my cancer surgeon, said that she'd handpicked him because she thought I'd enjoy him and that we'd connect.

Boy, was she right!

Later that night, while getting undressed to go to bed, I thought about my appointment with Dr. Ward. I stood in front of the mirror and looked at my breasts. This may have been the first time I *really* looked at them since I got sick. It was definitely the first time I'd examined them close up since my surgery.

The doctor was right.

My right breast was definitely smaller in size than my left.

Would I need more surgery to fix this?

I presumed I would at some point in time, or I could choose to live this way for the rest of my life.

And if the cancer didn't kill me, I hoped that would be a very, very long time.

At that moment, though, I couldn't bear the thought of one more surgery, especially on my boob.

When I awoke the next morning, I was stopped cold in my tracks by an article I spotted in the *Huffington Post* titled "My Strong Reaction to Joan Lunden's Bald *People* Magazine Cover Isn't What You Might Think."

In the article, the blogger, both a mommy and a TV producer, wrote about how her grandfather used to tell her that she looked like me when she was a little girl. At the time, she didn't get the comparison. Somewhere along the way, he had requested and received an autographed photo from me and had sent it to her.

While she was in high school, the blogger's father was diagnosed with inoperable brain cancer. He bravely battled cancer for just over three years. Ultimately, his brain cancer spread into his bones. He was unconscious on the day she graduated from high school, and he died less than a month later. Though it had been twenty years since her father died, when she saw that photo of me on the cover of *People,* she wrote, she was filled with hope—for me and for those battling cancer who would get to live the rest of their lives, happy and healthy, in honor of those who hadn't: in honor of her dad.

Well, you can bet I was floored and touched after reading this blog.

Her grandfather obviously saw something special in her at a very young age—and not because we allegedly looked alike but because there was a sparkle in her eyes and something that made her stand out in the crowd, I am sure. I was happy that she wrote and made the connection. And here we are, still connected, through her father's battle with cancer and his passing and now my cancer battle. So much has been learned about cancer since the nineties, when her dad struggled with it. I often think about what it must have been like back in the sixties for doctors such as my father. It must have been so challenging, so frustrating, so futile back then. With the amazing medical advances that have been made, many of us diagnosed with cancer today can fight it and live on.

A couple of days after my appointment with Dr. Ward, I decided to talk to Jeff and tell him there was something I hadn't shared with him in the meeting at her office. I explained to him that my right breast was no longer the same, and I wasn't sure how I felt about it.

Jeff took my hand and very sweetly said, "I love you no matter what. You don't have to worry about that. If you decide to do something about it, I will support you completely. And if you decide not to, it won't matter to me. I'm not a boob guy, anyway."

Still, I was worried about it. I knew it was something I'd eventually address. But I didn't know when or how.

What I was sure of was that the stress wasn't good for me.

The universe has a way of giving us exactly what we need when we need it the most. You have to pay attention and look for the signs. They're always there.

That afternoon, while I was contemplating so many things, Beth Bielat, my fitness trainer in Maine, sent me an email that said the following:

Remember,
Rest and reflect
Eat clean
Breathe and meditate
Exercise
And lots of family, support, friends and laughter,
Thinking of you,
    Beth

I am incredibly fortunate to have people who take the time to stop and send me important reminders in my life. I'm not sure I would have been affected by a simple email before my cancer journey, but I have a more profound appreciation now. And now that I do have gratitude for this type of support and outreach, it seems to come every day, in droves. It's the basic Law of Attraction: You get what you put out in the world. Every positive (or negative) event that happens to you is attracted by you.

Good morning Joan.

I've been watching you as you've shared your cancer journey. You've been brave, bold, and beautiful and so wonderfully open.

As a daughter of a two-time survivor, I salute you for arming women with critical information and empowering them to ask the right questions about breast cancer detection and treatment. And I salute you for your honesty in sharing the most intimate of your concerns and challenges along your journey.

With much admiration and respect,

Your longtime fan.

Paula Zahn

I was and continue to be so grateful for these emails from family, friends, colleagues, and strangers who feel like so much more, especially as I was on pins and needles going into my second round of chemo at yet another new place—though after I met everyone in the facility, I found them all to be incredibly nice.

On the morning of my treatment, Jeff and I made the fifteen-minute ride to the oncology center for my nine A.M. infusion. I will concede that, compared to the commute for my other chemo treatments, this short trip was a pleasure!

When we got settled, the oncology nurses opened up my port and checked my blood counts; it was business as usual. I was growing anxious and eager to know my counts, since my body had had well over a month to rebuild after the last round of chemo. I was expecting my counts to be up, or at least I was hoping they'd be.

When the results came back, I was a little disappointed. My white blood cell count was only 5.0; I had expected it to be back up around 7.0. The nurses explained that my body was still feeling the effects of the first bombardment of chemo, and the numbers were reflecting that treatment. The numbers they were more concerned about were my hemoglobin and my hematocrit, which were both alarmingly low. They advised me to take an iron pill and to be sure to eat red meats, dark green leafy veggies, and other foods that provide high sources of iron.

Once I was cleared to start, we got rolling. How much chemo you are given is dependent on your height and your weight. Together they are used to calculate your body surface area (BSA). Almost all chemotherapy drugs are dosed in this manner.

Based on my height and weight (no, I won't tell you!), I received two full syringes of bright red Adriamycin—which were not administered by the regular drip method but, rather, slowly "pushed" into the tube attached to my port.

(By the way, anyone going through this type of chemo should know that after receiving the Red Devil, you pee red for hours. I called it the "Hawaiian Punch Pee," because that's exactly what it looked like.)

While I sat in the chemo chair, hooked up to the tubes giving me my meds, I taped several videos for my breast cancer video blog. One of the videos I shot was thanking all of the women who sent me scrumptious hand-crocheted throws and scarves to keep me warm throughout my treatment. I was surrounded by many of them in the video, each one made by a woman from somewhere in America. It was so heartwarming that people took their time and put in so much effort to make these items for me. I thanked them all and posted the video on my website. I wanted them to know I was receiving their items and how much they meant to me.

Giving gratitude helped pass the time and filled my emotional bucket as the chemo dripped into my body.

I was given a lot of anti-nausea medications and steroids that first day. I usually left an infusion flying high from the steroids, and that day was no exception. In fact, the nurses wanted to give me their home addresses so I could go clean out a few closets for them. Those steroids hang around in your system for several days, which made those my favorite days of chemo—if there is such a thing, because once they leave your system, you crash and feel kind of like you have the flu.

When I got home later that day, I checked my email and found a lovely letter from a psychologist I knew through her son, who worked at Camp Takajo.

Hi Joan,

I saw your still, I'm very ashamed to say, enviably, beautiful, hairless face on *People* magazine and heard your incredible story on television. I don't know how you do it, but it looks as if you're thriving, not just surviving cancer. Obviously, the way that you really feel isn't in the scope of ordinary language. But you do manage to have an inexplicable gift that allows people to connect with you, not just watch. It seems you've learned to speak a language that's just as visceral as informative. And by some magical surge, you miraculously touched my heart and brain at the same time. So I wanted to write you a thank-you note of sorts. I'm sure that I'm just one of the many scared-of-cancer women who feels a little less petrified and has a lot more hope because of you. As a result of you daring so greatly, and allowing yourself to be publicly vulnerable, I have a sense of belonging to that tribe of worrier, warriors who are just a little less scared to join your battle against cancer. Needless to say, it's a spiritual connection. But it's there. Let me assure you that our entire family is fighting with you and sending you the most positive energy for a full and speedy recovery.

    With awe and hope,

    LC

I loved reading that note and truly appreciated the sentiment and connectivity LC felt. I was beginning to understand that my role was growing and my purpose was becoming more defined because of the impact of my willingness to share.

However, it was the following letter I received through my website on the same day that awakened something deep within me and helped me to understand that my true mission was something far bigger. This note really shook me to my core.

Oct 21, 8:29 p.m.

I just wanted to wish you the best in your fight against breast cancer, and tell you how proud I am of you. I also wanted to share some history. My grandmother, Violet Larson, was a 48-year breast cancer survivor when she passed away of natural causes on February 24, 2010, in Sacramento. I was truly blessed to have 45 years with my grandmother. And I really mean truly blessed, because it was your father who performed my grandmother's radical mastectomy. I am so thankful he was such a talented and gifted surgeon, because from what I was told, it was a difficult surgery. Bottom line, he saved her life and a handful of years later I was born and got to meet a fabulous woman! For many years in Sacramento she led the parade of survivors in the Komen Race. My mother is currently a 15-year breast cancer survivor. So far, I am good, but I know not to mess with family history, and I get my mammograms annually.

So good luck Joan Lunden with your continued fight!

Tracy M.

Okay . . . *that* was a big WOW!

I sat there motionless, with a tear running down my face, after reading this story about my dad. There was no way I could have imagined getting a message like that. There was a time, not so long ago, when it would have been impossible to connect with strangers on such an intimate level. I never would have known this story or the impact my father had had on Tracy's family without the access that social media has provided for people.

I absolutely loved that message and was so appreciative that Tracy took the time to share it with me. In the process of doing so, she shared a bit of my own family story. Reading Tracy's message conjured up a picture of me as a child, back at my home in Fair Oaks, California. I instantly imagined my dad in our living room, sitting in his big easy chair, surrounded by stacks and stacks of medical journals, which he devoured in his free time. I thought about how he would leave before dawn every morning, often before we awakened, because he had his early rounds and two or sometimes three surgeries a day. Although he had a medical office, he didn't spend much time there because he was always in surgery—at least that's how I remember him.

I can't begin to express how much I enjoy hearing stories about my dad. Since I was only thirteen when he was killed, I didn't have the chance to know him as an adult. Every now and then I hear from one of his old patients, which gives me a small glimpse into what he was like as a doctor and what he meant to people.

Wanting to know more about my dad a few years ago, I called upon three doctors in Sacramento whom he worked with and was close friends with until he died. I videotaped a sit-down interview with each of them. I learned so much about my father as a doctor, as a friend, and as a dad.

I often struggle to remember what his voice sounded like. I do remember that on weekends when he didn't have early surgery, he would be around for breakfast. He loved speaking Spanish with me at the kitchen table, since I was studying it in school; he spoke several languages but was rather fluent in Spanish. He was born in Sydney, Australia, was raised in China, moved to the U.S. in his teens, and often traveled to Mexico to deliver medical speeches at cancer conferences, in Spanish.

From his friends and colleagues, I learned that there were no "medical specialties" back when my dad was starting out in medicine in the forties and fifties, so he was what they called a general practitioner. When he was seeing patients, he could be doing anything from setting broken bones to delivering babies. Back in those days, an office visit was two dollars and house calls were five dollars. Dr. James Reece shared an office with Dad for years and told me that he was famous for not only writing a prescription at a house call but also for leaving a twenty-dollar bill under the prescription, knowing when a patient desperately needed the medicine but couldn't afford it.

I wish I'd had more time with my dad to know him better. I can't help but wonder what greater influence he would have had on my life.

With a growing practice, Dad recruited several other young doctors to his Sacramento office so he could concentrate on what he loved and did best—surgery.

I later learned from my dad's colleague Dr. Marvin Klein that he had a patient with a blockage in a major vein leading to her leg. The patient faced losing her leg or, worse, dying. There were no vascular surgeons in Sacramento, so my dad placed a call to renowned cardiac surgeon Dr. Michael DeBakey, at Baylor College of Medicine in Houston, Texas. Dr. DeBakey pioneered the first artificial heart in a human. Dad told Dr. DeBakey that he needed to learn how to perform this surgery to save a patient's life. He explained that he would fly himself down to Houston if DeBakey would agree to teach him to do the surgery. DeBakey happily agreed. My dad spent a week following him from one operating room to the next, learning about heart and vascular surgery, until he could return home to safely perform the needed surgery and save the patient's life.

I would never have known that story had I not set up those interviews with my dad's colleagues. Those tapes are among my most prized possessions.

By the time I was in first grade, my dad had become one of the top surgeons in our area. Several years later, he began to specialize in oncological cases. At the time, there were very few surgeons who specialized in cancer. He would sometimes fly to other cities to assist doctors on difficult cases. As a little girl, I remember standing with my mom and brother outside the big double doors of a hotel ballroom, watching my father at the podium to address a roomful of hundreds of doctors who were there to learn from him. I was too young to understand the importance of that back then. Only now, battling cancer myself, can I fully appreciate the significance of those moments and comprehend that my dad was one of the pioneers in the battle against cancer. He was out there in the world, valiantly trying to save lives with whatever methods they had available at that time. I have no doubt that he would be amazed to see the advances in the field of oncology. Perhaps he does see it all from somewhere beyond.

Gosh, I hope he does.

Simply imagining that he somehow watched me read the email from Tracy made me smile from ear to ear, if not gave me a good case of goose bumps. The fact is, Tracy would not have been born had my dad not saved her grandmother's life with that radical mastectomy.

There are no coincidences in life.

No accidents.

Sure, I knew my dad was a cancer surgeon, but somehow it never hit home that he was operating on women who had breast cancer—not until I read that letter.

Somehow I knew my father just a little bit better.

What a gift Tracy gave me.

I woke up the next morning slightly nauseated, and I immediately noticed that I felt fatigued.

This was the day I was told I would start to feel the real effects of the AC chemo.

Or maybe it would happen the next day.

Or maybe it wouldn't happen at all!

I spent a good part of my day resting—okay, sleeping—which was highly unusual for me. I was definitely having one of the crummy chemo days everyone had told me to expect, but it wasn't *that* bad.

I knew that I would lose my hair again, as the AC chemo attacks the hair follicles. Dr. Hollister had laughingly told me at the outset of this round of chemo not to fall in love with the peach fuzz that had grown back during the month since I'd ended the first round of chemo.

He was right; I had fallen in love with it. I'd asked Emir to come to my house to bleach that little bit of peach fuzz on my head so I'd once again be a blonde.

I know. I know.

It was all going to fall out, but until then I'd be a blonde, kinda, sorta.

After reading Tracy's email, I became a little obsessed with what I would find in my in-box each day.

I needed to keep things light, because I'd heard horror stories from other women about the AC chemo. I didn't want to feel sorry for myself.

I also heard success stories from people who wanted to cheer me up and inspire me.

Sometimes the letters were just to lend advice.

I always love getting advice, such as what I got below:

Date: Friday, October 24, 2014 5:06 p.m.
Subject: Ask Joan Submission
Name: Susan B

Question: I don't have a question, just a note from one warrior to another. I'm 54 and 4 years out from triple negative BC. I had a lumpectomy,

12 weeks Taxol, 4 rounds with the Red Devil, and 23 shots to keep my white count up. I had 7 weeks of radiation. What did I do? I danced almost every week. I love country music, so I kept dancing, just like I did before BC. Maybe not as much, but I was out there. I was a blonde. I was a brunette. I took a kayak lesson after I finished everything. I fought, because that's what warriors do.

Breast Cancer Awareness Month had been a dizzying adrenalin rush for me, but the real thrill was the reconnect with the American public that came with my appearances on the *Today* show. I spent days going through the messages and emails I received from thousands of people, I had such a desire and need to get back to as many as possible.

So many women.

So many stories.

So much sharing.

I took a look at Facebook to see what people were saying.

My name is Barb Miller, I am a single 50-year-old woman who lives in Belmar, NJ, and I was just diagnosed with breast cancer and I just read your story. God bless you and may your recovery go amazingly.

When I found out last week, I was in my car getting ready to leave Target, and I was told that I had breast cancer and I needed to come in right away for an MRI since the biopsy came back showing cancer. I have had my MRI and now tomorrow I sit with the doctor and find out what's what. You are so right, it happens so fast and it goes so quick and I am praying my survival skills and empowered-woman skills take over after I digest the news. I am not working and am single and my family has all passed away, so it is me and my dog Lucy and a few amazing women who will get me through this with stories shared by women like you. Yes, I can't even deal with the knowing of money and insurance issues right now and I have no clue of my path, chemo radiology lumpectomy etc., but when I hear it I will digest and do what I think is best for me and let silence and peace fill my soul. Thank you for sharing your story, I will follow along. Blessings, your friend Barb from Jersey

---

When I read a message like this, I felt like finding out where Barb lived in New Jersey, getting into my car, and going to her home to comfort her. I knew how difficult this battle was when you had a fabulous support system around you, so it was sad to hear from women who don't have a strong circle of support. And though I knew it wasn't possible, I wanted to be there for all of them.

Dear Joan,

I watched you on the *Today* show and you are truly an example of strength and encouragement. My breast cancer was in 2011 and ovarian in 2012. I carry the BRCA gene and would like to help others with breast and ovarian cancers. I am now two years NED (no evidence of disease—a term that you will come to cherish) . . .

The spotlight that you have courageously shined on breast cancer shows me that I can be an example to others too. If I can be of any help, I am happy to share the story of my journey with you. In the coming difficult weeks, I hope you have many good days.

Kind regards,

Wendy M.

Because of this message, I learned a new phrase that I was excited to add to my vocabulary. Wendy M. was right, NED was a term I would come to cherish. It's definitely on my bucket list!

In addition to the many well-wishers, some people wrote to give me useful tips and information. One gentleman seemed to know an awful lot about the digestive issues that occur when you are on toxic chemo, especially not being able to go to the bathroom. Some call it "chemo tush." I was very interested to hear his advice about having someone push on your back in a certain way to "get things moving again." He didn't explain how to go about pushing and where the spot was, so I wrote him back to find out. I was way too curious not to ask about this technique, even if it proved to be for my own personal amusement: *Hi Bill, I'm really curious about your ideas, especially at the moment, because I would love to "get things moving"—how exactly do you do that maneuver?* What I really felt like writing was *Hi Bill, I am totally full of shit . . .* Just kidding, of course, but if you didn't keep your sense of humor, then it all got to be a bit much!

And then, among the well-wishers, there's always someone out there who feels it's necessary to throw a dart at you, just because. I call them the haters. Often they have some agenda that inspires their caustic criticism; other times I assume they're just having a bad day themselves. Nevertheless, they are haters.

I'm Shannon and a 6-year (& counting) survivor.

I'm happy that you're kicking cancer's ass, however I feel that your touched-up photo leaves me . . . [I think Shannon was speechless because she forgot to finish her sentence!]

When normal women go thru this they usually don't have a full makeup team to pull them thru the hard days.

It's easy to be a role model, but even more inspiring . . . show the world, the real fighter! I wish you more luck than you'll need.

Shannon M.

Yep, that's Shannon M., and she's a hater.

Okay, let's be totally real for a moment . . . or should I say NORMAL!

How's that, Shannon?

Among the many lovely Facebook messages that I went through, doing my very best to answer as many as I could, there would always be those few haters in the crowd. Shannon M. thought I'd sold out on my bald *People* cover because I'd obviously had my makeup done.

Really?

I sometimes wanted to engage people like this, to try to kick them back and tell them I wasn't posting a bald shot on my Facebook page or taking some selfie, for Pete's sake, I was shooting a cover for a national magazine. They brought their top photographers and makeup artists, and guess what?

That's life.

Sorry if it offends you.

However, my daughters who work with me always have a hissy fit whenever I engage with anyone who isn't nice on social media, so I have to just let it go.

I did.

But then I went back and copied and pasted it so I could rant a little later.

I went on a pretty good tear in the privacy of my journal.

There's a lot of things I wanted to say and I got them out on paper.

You know what?

That felt pretty damn good!

Hey, you can't make everyone happy.

Chapter 27
My New Normal

The best side effect of fighting a life-threatening disease is learning how to live.

JOEL SIEGEL

Movie critic on Good Morning America, diagnosed *with colon cancer in 1997*

While I had finally gotten used to seeing myself wearing wigs, I can't honestly say that I would ever get used to looking in the mirror and seeing a woman with no hair, no eyebrows, and no eyelashes. It stopped me cold every time I caught a glimpse of *that* barely recognizable woman. It had been nearly nine months since I shaved my head, so in many ways, it had become my new normal, and yet there was nothing *normal* about it.

I sometimes looked into my children's eyes and wondered what they were thinking.

Really thinking.

I will admit, we didn't do a whole lot of talking about my disease.

My husband and I decided early on that he should be the one to do "Mommy updates," so the kids would feel free to ask questions about my health or prognosis that they might feel uncomfortable saying in front of me. For kids, the unknown is always scarier than the reality, so we wanted to give them that open forum. I recently asked Jeff if they were expressing any fears that he wasn't telling me about. He said he felt that kids were resilient and that this had become their new normal, too.

I wasn't sure how I felt about that. I didn't want Mommy being sick to be their normal any more than I wanted it to be mine.

That was compounded by the fact that the second round of chemo made it much more challenging to keep up my usual "I'm fine! None of you have to worry about me" facade. The toxic AC chemo dealt me a blow right out of the gate.

As it turned out, the toughest part of my battle was this poison—oops, I mean terrific medicine that killed the cancer cells that had taken up residence inside me but also brought me to my knees.

Once I started the second round of treatment, the phrase "chemo brain" became an often-used phrase around our house. I will blame every forgetful moment on chemo brain from now on—before and after the cancer.

Seriously, though, you are simply not yourself.

I also blame chemo brain for my inability to read longer than five minutes at a time.

While I'd planned on doing an incredible amount of reading when I was down for the count on chemo, I would lie down with my Kindle, so excited to pick up on a story that I was involved in, and within minutes, I would close my eyes. My eyelids grew so heavy, I simply couldn't keep them open. As much as I adore devouring good books, I didn't have the mental energy to read. So if you're going through chemo and you fall asleep while reading this book, I won't take it personally, I swear. I completely understand!

As my chemo progressed, I went out of my way to make my life very simple. I didn't tend to get dressed up, and I wore very little makeup. It's funny—when you dance with death, trivial things become less important. And somehow my face with a bald head and no eyebrows just didn't call for a lot of makeup.

Instead, I became totally focused on taking my supplements, what I was eating, getting in my work-outs, and rest. It was all about taking care of myself and winning this battle.

I couldn't wait until my *new* new normal became the old me.

After the buzz of the *People* cover wore off, I began to wonder when my hair would grow back and, when it did, what it would look and be like. I'd heard so many stories from people who had been through this: People with straight hair all of a sudden had curly hair, and sometimes it came back a completely different color. I was also told that due to the trauma, it could grow back silvery gray. Someone else told me that the first growth makes you look like a Chia Pet.

Gee, I could hardly wait for that.

So what would happen with mine?

Would it grow back curly?

I'd never had curly hair—wavy, yes.

I admit that I became oddly consumed by thoughts of when my hair would grow back.

Two months?

Six months?

I just wanted to look in the mirror and see myself again, *normal* Joan.

In the meantime, there was still a lot of work to do to get there.

Aside from the onset of a nasty cold, I thought I was doing pretty well.

Okay, wait.

I was actually quite worried about my cold. It was nothing but a normal lousy cold, but I was scared that I wouldn't have the immunity to fight it. I thought it would negatively affect my blood counts and jeopardize getting my next infusion. The longer I waited, the longer it would take to get through this round, and I just wanted the nightmare to end.

I was sitting at the desk in my home office, working at the computer, when I merely rubbed my right eye and a clump of eyelashes came off in my hand. I'd put mascara on my lashes that morning for the first time in ages because they had been growing back in so nicely. Although I'd lost my hair, eyebrows, and eyelashes once already, it's still shocking to see hair of any type fall out in clumps when it is supposed to be permanent on your body. I felt sick to my stomach, because this was another reminder that I was far from the finish line. I had three more rounds of chemo to go, and things were going to get harder before getting better.

While it's true that the second round of chemo was more physically challenging than the first, it wasn't nearly as bad as I'd thought it would be. Yes, there were some bad days, but I had prepared myself for much worse. It was manageable. Yes, I was greatly fatigued, and there were some uncomfortable digestive issues, but I'd had those with the first round, too, and over-the-counter remedies and eating clean had helped with that challenge.

The biggest difference between the first round and the second was mouth sores. While the chemo was killing off cancer cells, it was also killing off my good cells, especially cells on the inside of my mouth. When it does that, it changes your sense of taste, sometimes eliminating it completely. This can cause sores and inflamed taste buds on your tongue. At first I was only experiencing what I will refer to as "raised sores," which got better pretty fast.

Thank goodness!

However, because of my cold, whenever I put a cough drop in my mouth, it became instantly painful, so I'd have to remove it right away. And whenever I took a little DayQuil or NyQuil for relief, it was like putting battery acid in my mouth. Even Biotène, an oral rinse I was told to use to counteract and keep this condition at bay, became rather challenging; I quickly learned to tolerate that discomfort, because I knew the rinse would help and the pain would subside once I spit into the sink.

I started noticing that hot food and liquids were especially tough on my mouth. This was one of the scariest parts of chemo for me, because it made it hard to eat, drink, and feel normal.

If that continued, would it eventually make it hard for me to swallow?

That thought really scared me.

Joy, my oncology nurse, suggested that I try raw honey to help relieve the discomfort of the mouth sores, but she said it had to come from your locale. One of the nurses brought me a jar from a local health food market. When I used it for the first time, it definitely helped. Even so, the mouth sores became an ongoing battle that would continue and worsen throughout the AC round of chemo.

As I began my third AC chemo infusion, I realized I was officially halfway through my second round. I secretly pleaded with God not to let the second half of this round become much more challenging than it had been already. I knew it might be an unrealistic request, especially since chemo is cumulative, but I didn't think there was any harm in asking. The good news was that my white blood cell count was up to 8.0, which was terrific! With my blood count remaining consistently high, my treatment was working.

I told Dr. Hollister about the sores in my mouth. He said there was something they could do to help: Just before the nurses administered the "Adriamycin push," they would give me a cup of ice chips. If I sucked on the ice chips before the push, during the push, and five minutes or so after, the ice would constrict the blood vessels in my mouth. As the chemo rushed into my body, it wouldn't be able to go to the cells in the lining of my mouth because those vessels were closed. That simple act would prevent or at least help with the troublesome mouth sores.

Wow!

Who knew?

For some reason, I felt compelled to share with Dr. Hollister how lucky I considered myself because I was reading so many messages from women all over the country whose side effects were much worse than mine, making my physical journey look like a walk in the park. I also told him that I had recently done an appearance on *Access Hollywood*, where I received a couple of messages that stopped me in my tracks. While it was such a wonderful experience to hear from so many people, it was also tough, because sometimes they unintentionally put ideas into my head that I didn't have prior to reading their letters.

One of them said:

It sounds like you are almost done with your chemo which will mean that you will be declared cancer-free, cured, etc., soon. You will be a breast cancer survivor. While it is a joyful time for so many, it can be a time of anxiety for some as they contemplate the possibility of recurrence. Have you thought about having your cancer return, given especially that TNBC can cause a higher rate of recurrence? Do you have any insight on dealing with this fear? You have shown many how to cope with the challenges of breast cancer and maintain a positive attitude throughout. Any thoughts on recurrence fears after treatment has ended or when nearing the end of treatment?

Another said:

My friend was only cancer free for 5 months and had a recurrence, same breast . . . also triple negative. Had chemo, lumpectomy, and radiation. How do you get past the fear? I could only say it's in God's hands. Thank you in advance for your response!

Not only did these leave me asking, "What the heck do I say to *them*?"

But they also left me asking myself, "How *do* I deal with that fear?"

While the thought of "What if it comes back?" had crossed my mind, I hadn't given it a lot of weight—until that moment.

Dr. Hollister had a wonderful sense of humor and an incredibly infectious wit. Naturally, he also knew the right response. He looked at me and said, "I would turn to the great philosophical guru Wile E. Coyote—you know, the cartoon character who always ran off the cliff at top speed? Well, he was never in trouble until he looked down.

"Just don't look down.

"Keep your head up and expect the best.

"Don't go to the negative places.

"This AC you're going through and the radiation that will follow is like taking the morning-after pill. You take it and you don't get pregnant, but you also never know if you actually were pregnant or weren't pregnant, and you never will.

"This is exactly the same.

"You will never know whether you still had cancer cells in you or if you didn't, but at least you took that morning-after pill.

"As for the woman who asked about new medicines, let me just say that the oncology world is like a big ship with a tiny rudder. It takes a long time for it to change its direction. First it needs strong evidence that it should change; then it needs everyone in the profession to agree; and then some of the groups are still going to stick by 'how we always did it.'"

Goodness, he summed it up perfectly!

I was so relieved to hear his thoughts. And though I wouldn't have looked at Wile E. Coyote as a philosopher, Dr. Hollister had made a great point: Perception becomes reality. It really is how we see things that matters.

While I was getting my chemo that day, Lynn, the nurse navigator, came in to see if I had any questions about the next steps. I thought this might be a good time to ask about radiation; I assumed I should expect somewhere around four weeks in January. She said it might be four weeks or maybe six.

What?

Possibly six weeks?

That was news to me.

I had obligations and commitments already booked. I didn't have time for six weeks of radiation!

Lynn suggested I make an appointment with Dr. Ashwatha Narayana, the clinical director of radiation oncology and the radiologist who would be doing my treatments, so that he could look at my charts and make the decision sooner than later. She also said to make two immediate follow-up appointments, for a "simulation" and a "run-through," so that I would be ready to go right after the first of the year.

This was really good information and advice.

I went home and called the radiologist and got the first appointment they had. No one has ever accused me of standing still! Besides, you can't hit a moving target.

I was about to leave for my first radiation consultation with Dr. Narayana when Dr. Z called to suggest I speak to the doctor about scatter effects of radiation, which can lead to scarring of the lungs, blood vessel damage, atherosclerosis, and tissue burns. He wanted me to ask the radiologist what the transient and permanent side effects would be. As a cancer survivor, Dr. Z always wanted to make sure I was going into each part of my journey with another perspective and information in my arsenal when meeting my new doctors.

The first thing Dr. Narayana wanted to go over was all of the different choices of treatment I had.

Wait a minute.

More choices?

That meant more decisions!

Honestly, I just wanted him to point the damn radiation machine at my boob, blast me, and kill the cancer.

Apparently, it's not that simple.

Dr. Narayana explained how cells could grow out of control. He said my cancer cells had taken years to become the tumor that was discovered in June. There was the concern that other cells around the area might not have been part of that tumor and weren't considered cancer cells yet but were well on their way.

While chemo is used to kill the cancer cells, you also need to get rid of those cells that are on their way to becoming cancer cells, and that's where radiation comes in.

Dr. Narayana suggested that I weigh the good effects of radiation against the bad. Since my tumor was at the back of my breast, by my chest wall, I followed Dr. Z's lead and asked about the scatter effects of the radiation on my ribs, lungs, heart, and blood vessels.

Dr. Narayana explained that there had been a lot of changes in the way radiation was administered, so women didn't have as much damage to other organs in their body or as many long-term side effects. In the past, clinics had all women lie on their back with an arm up over their head during treatment; many clinics still do the treatment that way. However, research has shown that lying on your stomach on a special bed, with your arms over your head and an opening for your breast to hang down, allows the radiation to be directed so that it does not cross a path that would impact the ribs, lungs, or heart. Hearing this made the decision easy—I'd definitely lie facedown.

Dr. Narayana explained that they could direct the radiation specifically at the quadrant of the affected breast where the tumor was, or they could treat the entire breast. He felt that it was best to treat the entire breast and go after all the cells that might be in early stages of becoming cancerous. Okay—got it. I would go for the whole breast and get rid of all potential killer cells.

Next, the doctor said that the most reliable evidence showed it was best to go through six weeks of treatment. He already knew I was a bit taken aback by the prospect of having to go that long; there were some treatment centers doing a three-and-a-half-week treatment regimen. However, in order to do that, they gave higher doses of radiation, and that in and of itself could be tumor-producing. In his opinion, the best choice for me was a lower dose of radiation over a course of six weeks. This protocol would give me the best chance for a positive outcome and no recurrence. Since I had triple negative breast cancer, with a higher incidence of recurrence, this point was especially important for me to consider. I had to choose the six-week program if I wanted to give myself the best possibility of no more cancer.

So there really were no choices that day—just explanations of why I would go through this treatment facedown for six weeks.

I liked his approach.

I was in.

A couple of days before Thanksgiving, I went back to Dr. Narayana's office to get set up for my radiation treatments in January. They had me lie facedown on a special table that allowed my right breast to hang down in a hole so that it (and only it) could receive all of the radiation. They took measurements and photos of exactly how I was lying on the table and created a mold inside the pillow my head and arms were on, so that each time I went in for treatment, that mold would force me to lie the same way. The radiation machine wouldn't work unless all the measurements aligned perfectly, so this was an important part of the process. They also gave me three tattoos (just really tiny dots). The first was on the outside of my right breast; the second was on the inside of my right breast; and the third was on my back. Whenever I came back in, they would match up the lasers to the tattoos. That would keep the treatments consistent. I was mentally prepared for the tattooing to hurt, but it didn't. The dots were small and didn't go deep under the surface of the skin, so they were no big deal.

Does having three tattoos officially make me a badass?

Thanksgiving is always a day of reflection, but that year I couldn't help feeling I had something big to be thankful for: my life. As long as the cancer was really gone at the end of this treatment plan, I had a lot of gratitude to show and share.

I had no idea whether I was truly cancer-free. All I could do was hope, pray, and wait to see if anything came back.

While I knew this was a day to be thankful, I couldn't stop thinking about the woman who asked how I was dealing with the fear of my cancer returning after being deemed cancer-free, or the one who shared the term "NED and a survivor." Both really affected me. For the past week I'd been inexplicably teary, which wasn't normal for me. I couldn't explain what was triggering my reaction, but I would describe it as feeling raw and emotionally fragile all the time. Feeling like this was awkward, strange, and new.

Chapter 28
The Connection That Changed Me

There is no exercise better for your heart than reaching down and helping to lift people up.

My father

DR. ERLE BLUNDEN

Cancer surgeon

One of the most remarkable discoveries I've made on this journey has been the connection with thousands of people who have sent me their well wishes, prayers, advice, and stories of their own journeys. As in the quote that opens this chapter, I wanted to write about my experiences so that all the people who have reached out to me and given of their hearts understand that their small acts of kindness made a very big impact on me.

Although I didn't realize it, I had lost sight of how fortunate I was to hold such an amazing position on network TV and how that job connected me to so many people around the country in a unique and special way. I have come to understand it more fully over the past year as I have reconnected with so many of my "morning friends," and I have profoundly felt their love. I have also found immense strength and support in their messages. I suppose it sometimes takes a challenge to fully appreciate something intangible, like the intensity and breadth of a link to so many strangers far and wide.

If it weren't for social media, I never would have known people's capacity for compassion and kindness. I never would have been connected to so many Americans, heard all their amazing stories, their wonderful well wishes, or their helpful advice in my own time of need. I certainly never would have understood how much I mattered to so many—that they still cared so deeply, and with such kindheartedness.

On my last day of chemo, I woke up with what appeared to be another bad cold. I wasn't sure whether it was an actual cold, a sinus infection, or a side effect from the chemo treatments. The inside of my nose was terribly inflamed. I was coughing, sniffling, and my eyes were red and dry. I didn't want anyone to see me, let alone get out of bed.

But it was an important day—a significant day in my journey.

I was worn out, physically and emotionally, and I hadn't even stepped out of my bedroom door.

Jeff encouraged me to get up, but I wanted to wait until the kids left for school. I didn't want them to see me looking so worn down. I didn't want them worrying about me. I wanted them leaving that morning with their biggest concern being whether recess was indoors or outdoors.

For many, the last day of chemo is a day of celebration, something anticipated and embraced as a finale.

You know what?

I've never loved the way I feel at the end of a big party.

In fact, I always get post-party blues.

I'd been feeling emotionally charged that morning and in the days leading up to this last treatment—and not in a good way.

Throughout my entire journey, I had held on tightly to this warrior mode. But in doing so, I hadn't allowed myself to face a lot of my fears.

The ones I tucked down deep inside of me.

The ones I didn't want to face or talk about—with anyone.

The ones I didn't want to acknowledge or deal with or give one breath of life to—ever.

About a week before my last treatment, an oncology nurse took me aside and said, "Everyone reacts differently to the last day of chemo. Don't be surprised if you aren't totally elated. It's a mixed bag of emotions."

Thank goodness she gave me this heads-up, because I was emotionally all over the board. A part of me wanted to chalk up my mood swings to the medication and treatments, but another part of me knew that I'd spent the past seven months talking about little other than my disease, and now they were cutting me loose.

What did that mean?

The nurse told me that a lot of people got teary on their last day of chemo, while others were joyous and celebratory.

There was no right or wrong response.

That got me wondering: When is an ending *not* an ending . . . and would this journey ever *really* end?

The more I thought about it, the more I recognized that I was coming to the end of doing something about the deadly cancer cells in my body, but I wasn't done. Though I was coming to the end of chemo, I still had radiation to go through.

I think they call this winning the battle but still fighting the war.

So . . .

What was I supposed to do?

Sit back and wait to see if the cancer cells resurfaced?

The thought of being anything less than proactive was awful.

It made me terribly uncomfortable because I've always been one step ahead of everything. Except this, and even this I've tried to outrun.

This is the time when people start viewing you as a "survivor."

Was I?

Had I merely survived?

I wanted to do so much more with this new role I'd taken on. I wanted to continue my mission to spread awareness and make an impact. I wanted to keep changing women's lives. Every time I got an email thanking me for going public, for making women aware of dense breasts and ultrasounds, I had the potential to help save lives.

I wanted that to continue and become a part of my legacy.

And yet I was unbelievably and inexplicably overwhelmed with anxiety on the morning of my last chemo. And though I may have felt that way all summer long, I never—okay, rarely—showed it.

Whenever Dr. Weisberg asked how I was doing, no matter how bad I really felt, my answer was always "I'm doing great!"

She'd look at my blood work and then at me and say, "Oh, really?"

Somehow I overrode all of that.

But I knew I wouldn't get away with that on my last day.

And you know what?

I didn't really want to.

It was time to take down the wall.

Let the red cape fall.

Remove my mask.

Nobody can hold it in forever.

You can't be strong all the time.

Not even G.I. Joan.

When I walked in for my last treatment, the nurse in the room looked at me and said, "How are you today?"

I immediately burst into tears.

What with my system being completely worn down and feeling the need to be stoic throughout, holding in my emotions and doing my best to save face for everyone else's sake, my emotional release was inevitable.

I knew the outburst wasn't uncommon, but it didn't make me feel any better. That nurse found Joy, my oncology nurse, who was as demonstrative as I was that day.

"Damn you! I just dried my tears and blew my nose! I need to get a grip before Dr. Hollister gets here!" I said. I wanted to appear strong—pulled together.

I failed.

"It's okay. You're feeling it today. Don't worry," Dr. Hollister said in a very loving and kind way.

He went on to explain that my white blood cell count was good, but my hemoglobin and hematocrit were low. I was testing anemic. He didn't want me to do another transfusion. He preferred trying to build my strength back up through nutrition and supplements. He also reassured me that being anemic and run down from eight weeks of AC treatment would make me unusually teary and physically and emotionally worn down to the nub. He didn't want me to be embarrassed by how I was feeling.

"Think of this as the New Deal. Most people know that Roosevelt created the New Deal, but they don't know what it was or why it was called the New Deal. Everyone was physically and emotionally worn down from the war and didn't have enough money to do things or find jobs. Roosevelt created programs to put everyone back to work so they could put food on their tables and a roof over their heads. The reason they called it the New Deal was because he felt he was reshuffling the deck and handing everyone a new set of cards. They could take the cards from the new deal and do as much as they could with each one."

Then Dr. Hollister got very serious. He looked me right in the eyes and said, "Joan, you can't help the hand you are dealt, but you have the ability to play the right hand. You got dealt a hand that happened to have breast cancer in it, but you have played the hand as well as anyone could have possibly played it. You had every single one of your treatments. You had the latest, greatest, best medicine known to the medical community to treat your cancer. You took every one of your treatments right to the schedule, which very few people pull off, and the more you do that, the better your chances are for less recurrence. And now you've done this eight-week round of AC right on schedule, and I am sure you are going to do your radiation the same way. You have done everything right. You have a very aggressive form of cancer, and the chances of recurrence are much higher than others'. I am sure you know that. This has likely put fear into your head. I look at you and I say statistically, and not just statistics but my nose, and I have a good nose for this, you are cured. This is not to say I don't want you to do your six weeks of radiation, because I do, but in my opinion, I believe you've done it."

Hearing him say those three words—"*You . . . are . . . cured*"—meant so much to me.

A big part of my fear that day, which I didn't convey, was recurrence. As soon as you stop fighting the cancer in your body, all that's left to think about is whether one little cancer cell somewhere, somehow, slipped through. Was there one little cell waiting to show up or grow its own little blood vessels and become a tumor?

Fear had become my enemy.

I had to see it for what it really was in that moment: False . . . Evidence . . . Appearing . . . Real.

I needed to knock FEAR out of my thoughts.

I also needed to accept that if, for whatever reason, Dr. Hollister was wrong and my cancer came back, I would take care of it.

I wouldn't cave.

I wouldn't give in.

I wouldn't say, "That's it. I quit."

I'd fight, just like I fought this time around.

And if cancer came knocking on my door yet again, I'd kick cancer's ass all over.

I didn't need to live in fear or crumble under my angst.

The truth is, no one knows what the future holds.

All we can do is live in the here and now and walk our path one step at a time.

I felt so much better after my talk with Dr. Hollister.

It felt good to let my feelings out; to let it all go.

It was a really healthy thing to do.

It was like writing a new chapter in my book of life.

In a way, it released me.

By the time I got into the room to have my final infusion, I had no more makeup left on my face, and I was completely wiped out. Jeff, Lindsay, Sarah, and George's mom, Rita, were all waiting for me.

"What took you so long?" someone asked.

I simply didn't have the interest, strength, or energy to answer.

During my last infusion, I posted a photo on Facebook with the short message: *My last chemo, I can't believe it's over.*

When all was said and done, the nurses gathered around and gave me gifts and cards, congratulating me on my triumph.

I didn't feel like a victor, though I did my best to smile and take it all in, thanking everyone for their love, guidance, and support throughout those eight horrific weeks. The work that those nurses do—that *all* nurses do—the compassion, the care, and the drive that gets them out of bed each morning is awe-inspiring. They face death every day. They deal with cancer every day. They deal with chronic illness every day. They are the true victors in this story: the real heroes to be celebrated.

Much to my surprise, later that day Sarah called to say that my eight-word Facebook post had received an amazing reaction! By seven P.M., the post had reached 1,104,384 people and been liked by almost 80,000. Nearly 4,000 well-wishers had left wonderful comments.

When I saw the amazing response on social media, it truly took my breath away.

But it didn't stop there.

The numbers kept going up and up and up.

A few days later, my post had reached almost 8 million people. It had been liked by almost 800,000, and 18,629 comments had been left, mostly messages of encouragement that came in many forms.

That boggled my mind for many reasons, but mostly because I could hardly comprehend that so many people cared. I know I've said this before, but my journey was ongoing. And yet people kept up with it. That was amazing to me. I was heartened to post something that elicited so many warm, encouraging, thoughtful comments from so many, but especially from millions of people I didn't know. Not only did it boost my mood, but it boosted my faith in humanity. Above all, it gave me great hope for the future.

This journey has been nothing short of extraordinary for me. While it has been a roller coaster full of twists and turns that were both highly emotionally charged and exhausting, I must say it was made much more palatable when I saw some of the responses to my "it's over" post.

Why?

It confirmed that I was moving toward a more meaningful and purposeful life that was inspiring others on a deeper level.

By the time I awoke the next day, I was feeling relieved—like a boulder had been lifted off my shoulders. I was feeling better physically, too. I quickly rolled over in bed to check Facebook for more messages.

I was *shocked* to be greeted by five thousand new comments, one of which was from a forty-year cancer survivor who wrote, *Thank God. Live each day as a gift from above. LIVE—LAUGH—LOVE.*

What a wonderful way to greet the brand-new day.

And a new day it was, indeed.

Chapter 29
Unbroken

*I do not feel any less of a woman. I feel empow-
ered that I made a strong choice that in no way
diminishes my femininity.*

ANGELINA JOLIE

*Actress, filmmaker, humanitarian, tested positive
for BRCA 1 in 2013*

After being on maternity leave for three months,
Lindsay officially came back to work in early
December. Sarah had done a fantastic job filling in for
her sister and really helped save the day in her absence.

I don't think any of us anticipated the impact my
breast cancer was going to have on my office. I wasn't

physically in the office all summer. To be fair, that wasn't unusual, since I usually spend the summers in Maine, but I had not been interfacing with them like I normally did. Once I'd returned to Connecticut in the last week of August, I hadn't gone back to my usual working schedule. I'd been mostly phoning it in from home. Meanwhile, Elaine, my personal assistant who was in charge of every event and appearance, was juggling my schedule from the office. She was helping me assess what I could still do during my treatments, and she made sure to keep in touch with every speakers' bureau, so that when my treatment was over, they would know I was ready to go full speed ahead. Both of us were somewhat surprised by the number of requests I had received to speak at various cancer events around the country. It was wonderful but also challenging to decide how many I could agree to do, especially in the midst of my treatments. Thankfully, Elaine was extremely diligent about every detail expected of me, so we could digest the plethora of offers and make decisions without jeopardizing my health.

By the end of the year, I wanted everyone in my office to fully understand what I'd agreed to do in the coming year and how 2015 would be unfolding. Many of the speeches would be based on my breast cancer journey, which I found exciting and was looking forward to sharing with others in both cancer research and similar situations.

I will admit that it felt good being back in the office. It was so natural and comfortable having the whole team back together for the day.

Secretly, I couldn't wait to get all the toxic chemicals out of my system so I could shake the lousy side effects I'd been feeling. I'd had a constant runny nose, a hacking cough (which I'd discovered was from the Cytoxan—the "C" in my AC chemo), and the inside of my mouth was still terribly sore. I'd also made an appointment the following week to finally have my chest port removed. I just wanted that thing out of me. It was one more reminder that screamed "CANCER PATIENT" every time I looked down at it.

For a couple of weeks, whenever I turned on the television, I'd catch glimpses of the trailer for the new movie *Unbroken*, directed by Angelina Jolie. Whenever I heard the line "If you can take it, you can make it," it resonated deep within me. It summed up exactly how I had been feeling for months. As luck would have it, Jeff and I were invited to a private screening of the film in New York. Just as the movie begins, the single word "Unbroken" appears on the screen. When he saw it, Jeff slipped his arms around me in the darkened theater and whispered sweetly into my ear, "That's you, baby, *unbroken*."

After the showing, we were invited to the Porter House Restaurant in Manhattan for a private reception where Angelina Jolie was expected to attend. I wasn't sure we would get to see Angelina Jolie, let alone have a chance to speak with her. However, as we made our way through the restaurant, all of a sudden Angelina appeared before me. She is absolutely breathtaking, lovely and truly gracious. I was compelled to approach her. Not because she is a movie star but because I admire her bravery in coming forward with her story and telling the world about the BRCA 1 gene. She exposed her battle and, I am sure, saved many lives in the process. When we met, we spoke about the film but also about breast cancer. Although our journeys have taken us along different paths, we found that we had much in common. We also swapped stories about having so many kids.

Jeff said to her, "What's another couple of hot dogs on the grill?"

She said, "That's what Brad always says!"

Jeff and I talked about this exchange for days.

I thought I had been doing better, feeling well, but a few days later, I was stunned to awaken feeling horrible. It was, without a doubt, my worst day since starting chemo. I ached all over. While I would love to say it felt like a terrible case of the flu, it was so much worse.

I spent the entire day in bed, sleeping, trying to make my angst pass. By midafternoon my temperature had spiked to 101. I needed to call my oncologist. That was the rule—if my temperature reached 101, I had two choices: I went to the ER or I called the oncologist.

Dr. Hollister was in San Antonio at a breast cancer conference so I spoke with another oncologist who was on call, whom I'd never met. She prescribed an antibiotic to protect me, since my white blood cells couldn't. She reminded me to take Tylenol to help bring my temp down. If I didn't improve, I would need to go to the ER. I was completely listless. I had no energy.

Although my temperature dropped, my condition continued for the next several days. I couldn't get out of bed. When I finally did, I was dragging, big-time. These were unlike any days I'd had. I was moving at a pace that was unrecognizable. My older daughters had gathered around me, but I wasn't engaging. They would give me something work-related, and I would become so overwhelmed by the prospect of having to make a decision that it reduced me to tears.

There was no way I could go in to have my port removed, as scheduled. I was too weak and sick. My doctor wanted me to get a chest X-ray to make sure my hacking cough wasn't turning into pneumonia. They also wanted a complete blood test to see what was going on.

Thankfully, the chest X-ray was clear, but my blood results showed that my white blood cell count was astonishingly low: 0.6! My hemoglobin and my platelets were also really low.

These numbers put everyone in alarm mode.

The cumulative effects of my chemo had finally caught up with me.

Did I need another blood transfusion?

I was really scared.

I just felt so physically vulnerable. My doctors were stressing the importance of staying hydrated, especially with a fever. Everyone was telling me this, but water tasted so bad to me, like metal. I also had no real taste for anything, so nothing sounded remotely good to eat or drink.

These were, by far, my darkest days. All I wanted was some relief.

After a couple more days of feeling downright awful, I called Dr. Oratz, whom I considered my cancer quarterback. She said, "You don't have to feel this lousy. We can help you. We are going to give you intravenous fluids, a blood transfusion, and then a shot that is one of the complementary treatments to chemo that helps make more white blood cells." While this meant my port would be staying in for a while, I didn't care. I just wanted to feel better.

After the fluids, transfusion, and shot, I began feeling normal. I couldn't believe the difference in how I felt after being so sick. I actually felt human again. What a difference the blood transfusion made!

As the holidays were approaching, Jeff and I were preparing to take our family to California for two weeks of some much-needed rest and relaxation. We had rented a home in La Quinta, near Palm Springs. I was really looking forward to having some time in the warm sun. Unfortunately, the normally warm desert was experiencing a fluke cold spell. So much for global warming!

I was still experiencing such extreme fatigue that I spent most of the week in a relaxing recliner inside the house, playing Scrabble with Sarah and her friend Stacy and watching reruns of *NCIS*. I'd planned to catch up on my reading, but I was still finding it difficult to focus on and stay with a book for any length of time. I was beginning to wonder if I just needed a better or more intriguing book, or was the fatigue still that overwhelming?

Palm Springs proved to be a difficult trip for me. No matter how hard I tried to be outside and playful, I was always freezing cold and perpetually exhausted. I just couldn't seem to catch a break. However, Kate and Max and Kim and Jack managed to take advantage of the desert vacation. Jeff had also invited his parents and his sister, Karen, who has three children, Kenny, Keith, and Kendall, who are similar ages to our kids. We had a full house!

The kids took a few intrepid dips in the pool when the sun poked out, and they played football incessantly in the enormous backyard. I was happy that they were having fun—that is, until one evening, when they were playing "two-hand touch" football. My youngest son, Jack, was running fast to catch a long pass. Looking over his shoulder at the ball and not where he was going, he ran smack into a palm tree, leaving a nasty bruise and a huge bump on his temple. Jeff and I rushed him to a local emergency care center. After sitting for close to an hour in the waiting room, I ended up having to wait in the car after a baby threw up next to me. Our first thought was "And what is that baby here for? Hopefully nothing contagious!" The last thing I needed in my already weakened immune state was to contract some other illness. Needless to say, it wasn't the vacation I was hoping for, but I did my best to hang in there and be present for my family.

Truthfully, I was anxious to get back to Connecticut and start my radiation treatments. As soon as I could do that, I would feel closer to being cured.

My required dose of radiation called for treatment every Monday through Friday for six weeks. At the outset, that seemed like such a long time and a very big commitment, but there are no shortcuts in the battle against cancer. I was willing to do whatever it took to get to the finish line.

By the middle of February, I had crossed that finish line—or had I? I had finished the last day of my five-week regimen of photon radiation, during which they shot radioactive beams through my boobs. My right breast was deep bright red, as if I'd gotten dropped on a desert island wearing a wet suit that covered my whole body except my right boob, which was exposed to the bright island sun day after day. My breast was inflamed and tender to the touch. Some spots were flirting with parting and becoming open sores. I was slathering on thick skin cream made especially for burns four or five times a day for relief, then covering the wounds with bandages.

I had one last five-day regimen of "electron radiation" to go. I would be lying on my back for this round, with my arm up over my head. It was bittersweet to be ending in the same position I'd started. I'd gotten over being embarrassed about all of the technicians, male and female, walking around looking at my exposed boob, making sure it was in just the right position for the beam to penetrate down to the cavity where the tumors *used to be.*

Gosh, that felt really good to say.

Chapter 30
Had I Known

The goal is to live a full productive life, even with all that ambiguity. No matter what happens, whether the cancer never flares up again or whether you die, the important thing is the days that you have had, you will have lived.

GILDA RADNER

Comedian, diagnosed with ovarian cancer in 1986

When I first heard the words "You have cancer," deep down, I knew exactly what that meant. It meant *CANCER:* the BIG "C."

It also meant this was the real deal, and yes, I could die from it.

But then I started the battle, the fierce fight against the cancer cells, and the warrior in me hoped I had what it would take to win—that when all of my treatments were over (the chemo, the surgery, and the radiation), my disease would be gone for good, never to return.

I remember reading a social media message one day that said, "Hang in there. Before you know it, you will be looking at this cancer through the rearview mirror." It was an incredibly uplifting thought. I knew or at least hoped that someday I would be able to look at my cancer that way—through my rearview mirror. And in time I was certain I'd get there.

The overwhelming question was how much time *did* I need to actually "survive" before I could begin to feel like I'd truly beaten it?

Don't get me wrong. There were days when I doubted my chances, but I also didn't want to set myself up as some model "super-survivor" who never looked back, never worried, and somehow faced each day as an invincible trouper.

Was that even possible?

When I was growing up, my mom taught me three great life lessons I've carried with me:

First—always be positive.

Second—expect great things out of life.

And third—whenever things get tough, always expect a better tomorrow. Mom was not born to a rich family and lived an extremely humble life. When she finally found love and security with my father, it was dramatically ripped away from her when he died. If anyone had a reason for being somewhat leery of life, she did. And yet her positive attitude was so effusive, it was downright contagious. So believe me, I learned from the best.

Over the course of my treatment, I read countless tweets and Facebook messages from people who encouraged me daily to "stay positive." And not give up the fight. They were important in my recovery because I felt a bond and connection with those who had walked the path before me, those who knew where I was headed and could hold my hand and lead me through the darkest days with their words, observations, and experiences. Out of everything I received, the overwhelmingly consistent message was how important maintaining a positive attitude is when you are battling a disease. But only those who have gone through the battle against cancer really know how hard and scary it is! You just have to remember not to let a bad day make you feel like you have a bad life.

However, cancer is a powerful and daunting opponent.

Cancer looks you in the eye like a big angry bull with huge pointy horns, ferociously breathing as though he's ready to attack and head-butt you into some other stratosphere at any moment.

You would think that the further along I got in my cancer treatments, the more chemo I'd sent pulsing through my body, and the more radiation I allowed into my system, the more relieved I would be, right?

As if I stared that damn bull down, and at some point he went running off in the other direction and I won.

Ironically, for many, including me, the closer to the end of our treatments we get, the more worried we become about whether we truly beat the cancer.

And when you are completely finished with all of your treatments and everyone is cheering you, "You're done, you beat it, and it's all over!," you want to join them in that joy, believe that it's all over, but it takes a while to embrace that.

And that's okay.

Every one of us who has battled cancer lives with the knowledge that it could return. Carly Simon wrote an entire album about having the chemo blues, the depression that followed her breast cancer treatment. I can tell you that as hard as you try to be reasonable, it is difficult not to experience some anxiety and to think:

Did we really get it all, or am I waiting for the other shoe to drop?

Of course, being the research junkie that I am, I googled "What is the effect of worry and stress on cancer growth?" Research indicates that stress can alter the immune system function. In turn, immune system function can alter tumor growth and response. Obviously, worrying wasn't going to do me any good!

The key to being a survivor is not letting that fear overwhelm you. You must find the strength and courage within yourself to let go of the fear and enjoy life; otherwise, the monumental battle you just fought to overcome your cancer will have been for nothing.

I had to remember not to allow myself to waste precious time.

Time fearing a recurrence, because that may not happen.

Time with loved ones, because they matter more to me today than ever.

Time doing things I'm passionate about, because what's the point of doing things you don't love?

Time taking care of myself, because my health matters.

Time focused on things I can do something about and not on things I can't change.

Time enjoying the moments because they pass so quickly.

Time acting on things NOW, because tomorrow might be too late.

Time . . . it is so very precious.

In terms of my entire life, this nine-month battle against breast cancer has been a blip on the radar screen for me. As daunting as my fight has been, it did not break me. Not even close. I am grateful to still have that mind-set. Being able to see my disease from that point of view helped me change my focus, as if it were a lens on a camera that helped me go from a close-up view— which we all know makes things look proportionately bigger—to a panoramic view, which ultimately helps put everything into greater perspective.

When faced with such a threat against your mere existence, you do begin to appreciate life so much more. You look at each and every day through that new lens. As I looked out my window from my lakeside home last summer in Maine, I saw the sun glistening off the lake. It was a more gorgeous sight than I remembered it being in past years. In fact, I seemed to appreciate every detail of that view more, just as I do my life these days.

If there is one, *that* is the silver lining of cancer.

It certainly makes you more appreciative of life.

A woman on Facebook wrote to me about this epiphany: *A wise friend of mine explained it all like being given a present wrapped in barbed wire!*

How true!

Joyfully, breast cancer wasn't a death sentence for me. In a bigger way, it was a a wake-up call—one that taught me some very valuable life lessons. I never could have dreamed I'd learn so much from this challenge, but like they say, *that which does not kill you makes you stronger.*

In thinking about my future, I couldn't help but wonder if, after having breast cancer and questioning my very existence, I would find myself rushing to do the things on my so-called bucket list before it was too late.

In retrospect, *had I known* that breast cancer was going to have such a powerful and positive impact on me, would I have made some of these changes earlier in my life so I could have enjoyed the benefits of the outcome sooner?

That's the million-dollar question—one I suppose I'll never know the answer to.

But here's what I do know.

While I desperately want to put this ordeal behind me and live life as it was BBC—no, not the British Broadcasting Company, life *before breast cancer,* life when I didn't worry about whether I would be around tomorrow—I don't think that will ever be possible again.

And in many ways, I am appreciative for that gift.

We can't go back and change what was.

If we're given the chance, all any of us can do is move forward with our lives and hope to do better, be better, love better, and live better.

These are choices that are within our reach.

I'm not going to take that for granted.

So, as I move forward, I don't want to go about my life as if nothing happened.

The reality is, something did.

Something seismic.

As a result, there has been a tremendous shift in how I see, feel, touch, taste, and take in . . . well . . . just about everything.

Yeah, a second chance at life will do that to you.

With that in mind, my old bucket list seemed a bit insignificant. It was lacking purpose, meaning, and direction. So I thought it might be a good idea to make a new bucket list.

The first few things I wrote down were:

1. Being NED (no evidence of disease).
2. Continuing to have clean nodes.
3. Grow a full head of hair.

4. Healthy blood counts.

5. Clear ultrasounds.

6. Eat clean, healthy foods—hopefully in exotic faraway places as often as possible!

7. Write more books, sharing my journey and what I've learned along the way.

8. Inspire. I can't believe I once had a real fear of public speaking. Now I live for it.

9. Hike more. I don't really care where. Just as long as I'm forging my way through nature.

10. See my older daughters create their families and live out their lives *happy and healthy.*

11. See my younger kids graduate college, marry, and have babies of their own—in that order.

12. Play better tennis and maybe learn to golf when I'm a senior—in twenty years or so.

13. Make an impactful difference in the life of others.

14. Save lives—I know, I'm still a wannabe doctor— but curiously, I might.

15. Grow old . . . with Jeff.

16. Nothing—I'd like to be comfortable just doing nothing at some point in my future.

What's on your bucket list?

I think there's great value in creating a bucket list, one that will take a long—*long*—time to complete. One that will inspire you and give you purpose and give you momentum and hope.

One that spans a lifetime.

We've all gone through something in life that has changed us in some significant way, and as a result, we will never go back to the person we once were.

Life takes each of us down different paths, and sometimes those paths actually change you as a person.

As 2014 came to a close, I received an email from Robin Roberts, who was checking in on me and wishing me a happy holiday season. Normally, I send out holiday cards to my family, friends, and colleagues that are clever and campy pictorial overviews of the year. The card captures moments of each family member— and in 2014, all *twelve* of us, the kids, sons-in-law, and baby Parker Leigh, together as a family. It was no easy feat getting my cards out like I usually do, because I was feeling so lousy in early December. Somehow we managed, and despite the wide swing of the pendulum in 2014, when I looked at our card, I had no doubt that there was indeed far more to celebrate than not.

One of the greatest moments of joy I received to close out the year was the blessed news that my oldest, Jamie, and her husband, George, were expecting their first child. Getting pregnant hadn't been an easy road for them, so when they shared that they were indeed expecting, I was thrilled beyond words. My heart was once again full of deep bliss and, yes, peace. You see, despite all of the trials and tribulations I had gone through, life does go on. My family—my children—were all thriving. They were blossoming and doing the things that filled their emotional buckets, creating families of their own that would endure and carry on the legacy I had instilled in each of them. This thought alone gave me satisfaction, fulfillment, and total contentment that I know any mother would surely understand.

So when I wrote back to Robin, I was feeling especially elated and full of gratitude. I was also a bit wistful and more reflective than I was in our usual correspondence. While I couldn't share the most exciting news about Jamie and George with her—it was too soon to go public—I wanted her to know the other ways this journey had touched my heart, because I knew if anyone would understand the connection and change I felt, it would be Robin.

I shared with her that the most wonderful and surprising part of getting cancer was that it had reunited me with so many Americans who had spent mornings with me for decades and now, through my public cancer journey, have reconnected with me through social media.

This reconnection has been another surprising silver lining for me. One would not set out on this journey expecting it to bring happiness, but for me, it did. It has been an unexpected blessing to touch so many women's lives. I've received messages of thanks and gratitude from countless women telling me that they had never done a self-exam or hadn't had a mammogram in years, and because of my speaking out, they got checked. Because of that, many actually discovered they had cancer, but we know detection and treatment will save their lives. For that reason alone, 2014 was a great year!

In February 2015, I was asked to be the keynote speaker at the largest and longest-running breast cancer conference in the United States—the Miami Breast Cancer Conference—attended by well over a thousand medical specialists who diagnose and treat women with breast cancer. The attendees are primarily breast cancer surgeons, oncologists, radiation oncologists, radiologists, and breast cancer support team members. The physicians hail from all fifty states, Canada, Central and South America, and Western Europe. The meeting has a significant impact on practice standards, dissemination of information, and research directions for breast cancer in the United States. My invitation to speak came directly from Dr. Patrick Borgen, the chairman of the conference, who said he wanted me to talk about the "all-important piece of the puzzle" for doctors: the patient's perspective.

I was not only honored by the invitation, I was also a bit taken aback by the distinction and the opportunity to address such a prestigious group of professionals. This would be the first time I'd address a large group of important medical professionals on what it was like to go through treatment from a patient's experience— and to share my personal journey in a public forum. I felt I had a rare opportunity to enlighten these medical professionals from an "inside" point of view. As both a journalist and a patient, I was pretty sure my insights would be thought-provoking; however, my real goal was to be the voice for the thousands of other women just like me.

Preparing for the speech was extremely nerve-wracking. I was actually scared about what they might think. I had been given a rare chance to address an important audience on a topic I had become passionate about. I needed to thoughtfully construct my message so they really heard what I had to say.

And boy, did I have some things to share.

I wanted the doctors in that room to understand that when women across the country first heard my story—that I initially had a clean mammogram, and ten minutes later, an ultrasound found an aggressive cancer tumor—they were stunned. Many of those women had no idea that vital information wasn't being passed on to them, information that could risk their health. I felt it was critical that women be given every available resource to make the best health care decision. When women are informed that they have dense tissue, only then can they make potentially lifesaving choices about their care, in consultation with their doctors.

Frankly, it was just pure dumb luck that I interviewed Dr. Susan Love and got that ultrasound in the first place. *Had I known* about dense breasts, and that my mammograms wouldn't show a tumor if I had one, then I would have treated my breast screenings with a very different perspective all those years. *Had I known* that there was lifesaving information not being divulged by my mammograms and not being passed on to me, the patient, I would have been outraged much sooner and probably would have been out there blazing a trail to have that fixed for women everywhere. Then again, maybe you can only effect that kind of change if you've been diagnosed with the disease yourself.

But now I belong to that sorority. I can stand up and do everything I can to be heard and to make a difference. And the Miami Breast Cancer Conference was my first significant opportunity to prove I could do that.

The American Medical Association's code of medical ethics states: "Withholding medical information from patients without their knowledge or consent is ethically unacceptable. The assessment of the potential benefits and harms of a specific test or treatment can be made rationally if the information given to the patient is complete, accurate, and true."

We know early detection is important to surviving breast cancer. However, despite the fact that 40 percent of women have dense tissue, survey data from Harris Interactive poll a few years ago indicates that 95 percent of women do not know their breast density—which can change over the course of their life—or that it matters.

It does matter.

Had I known, it certainly would have mattered to me.

Not long ago, I had an opportunity to speak with Dr. Nancy Cappello, whose late-stage cancer was found a few months after a clean mammogram; fortunately, she is a survivor and a fighter. She is the founder and director of Are You Dense?

As a result of the incredible passion and advocacy of women like Nancy, we now have twenty-two states with legislation requiring doctors to inform their patients that their breasts are dense and that the mammogram may not be a sufficient screening tool. I look forward to working hard to help her efforts to bring all states on board.

I am also working with Senator Dianne Feinstein on legislation to make a federal standard that every radiology lab in every state needs to tell every affected woman in real, simple language that she has dense breasts and that she may require further testing.

So there is progress being made in the responsible reporting of dense breasts and also in women's awareness of their breast health.

During the writing of this book, my coauthor went for her annual mammogram. For the first time ever, she saw a boldly printed sign in the waiting area of her doctor's office:

WHEN YOUR MAMMOGRAPHY IS PERFORMED, ASIDE FROM CHECKING FOR ANY CHANGES OR ABNORMALITIES, WE ARE CHECKING HOW DENSE YOUR BREAST TISSUE IS. THIS IS A NYS MANDATE. IF YOU ARE 75% DENSE, A SCREENING ULTRASOUND WILL BE RECOMMENDED. PLEASE FEEL FREE TO ASK ANY QUESTIONS.

Coincidence?

I've never been a big believer in those.

As they were placing her breast on the metal plate, she mentioned that she was curious to know if she had dense breasts. With that, she heard in an exasperated tone: "Oh, sure, like every other woman who comes in here now, ever since that Joan Lunden was on the cover of *People* magazine."

Apparently, this was a request she was getting *a lot*!

My coauthor waited to break the news about her connection to me until *after* her breasts were out of the vise. Probably a good idea.

You might think that this is where I'd feel the necessity to issue an apology to everyone involved in radiology, but I will not.

Nope.

I refuse to apologize for informing and empowering women to better understand their breast health and advocate for their care.

On the contrary, I feel like I must boldly step forward and question the special task force convened in 1984 to study whether to continue recommending that mammograms begin at forty. The U.S. Preventive Services Task Force has issued mammogram guidelines recommending that women begin mammogram screening at age fifty and repeat the test every two years. This is heatedly debated in the medical community. The American Cancer Society and most other major cancer organizations still recommend that screening begin at forty and that mammograms continue annually. The task force was formed after a small Canadian study brought into question whether starting mammograms at forty might cause too many false positives (and unnecessary biopsies) and that not enough lives were saved to justify the testing.

Really?

How many lives would need to be saved for this panel to say they're worth it? I don't know about you, but I'd rather have a false positive any day, rather than a false negative, the way I did with my mammogram.

Call me crazy.

I'd like to hear from the thousands of women around the county in their forties who are alive today because they got a mammogram and, as a result, got treated and lived to tell about it. Are they just not worth it to the task force?

Maybe we need one of those women on the task force.

The doctors at the Miami Breast Cancer Conference were all polled on this issue. I stood at the back of the gigantic ballroom where I would be speaking to them later. I watched as each of them voted on this question, and it was close to unanimous to leave the standard age at forty and the testing on an annual basis, although it should be individualized with each woman and her risk factors.

Several doctors got up and expressed concern that the insurance industry might be able to argue that the underlying expenses would support the delay of screening until age fifty, and how terrible that would be for women everywhere.

It's time to wake up!

We need early screening and, when necessary, ancillary screenings to be made available to every woman and covered by insurance.

Just consider the facts: We are seeing more and more women in their twenties, thirties, and certainly forties being diagnosed with breast cancer, and researchers are frantically trying to figure out why.

Doesn't this clearly show that early screening is more critical than ever?

As much as this cancer has tried to beat me down, somehow it has also grown me a big ol' set of balls. I'll be careful not to wear my skirts too short when I go to Washington to advocate, so they won't show. But as long as I grew them, I might as well use them.

There's no question that we've made great strides in our fight against breast cancer, but my experience, combined with those of so many other women diagnosed with the disease, shows that we still have a lot of work to do. Many people have worked tirelessly to raise awareness, to spread the word that breast cancer is a threat to women everywhere, and to suggest that funds are needed for more research. As a result, we have achieved great awareness around the country about breast cancer.

When the NFL is wearing pink during October, I think we've got plenty of awareness.

What I believe we need more of now is *education*.

Thankfully, millions of women are now receiving routine mammograms, but going forward, I hope that we shift the focus to better tailor health care screenings to fit each woman's individual needs. *Breast cancer care cannot be one-size-fits-all.*

It is a very heterogeneous disease. It is a cancer that I believe requires great personalization. What's right for one woman isn't necessarily right for another, and approaching it as a one-size-fits-all issue undermines the true abilities that we have to combat breast cancer.

I hope there will be a day when we can get gynecologists and other referring doctors to discuss the issue of breast density with their female patients. Women have been kept in the dark on this issue far too long.

Had I known that my radiologist had been giving lifesaving information to my gynecologist for decades that wasn't being passed on to me, I would have been able to have an intelligent conversation about my risk factors. Instead, I stuck my head in the sand, thinking and believing that because I didn't have a family history of breast cancer, I was somehow immune.

Had I known that dense breast tissue increases your risk of cancer, maybe I would have been more vigilant in my care. I hope that as we go forward, we'll focus research not only on treatment but on *prevention* and which lifestyle factors and habits might alter breast cancer risk and recurrence—exercise, weight, diet, and stress.

Had I known how bad sugar is for you—that it is like jet fuel for the growth of cancer cells—hello, I would have cut it out of my diet a long time ago. It's sugar's relationship to higher insulin levels and related growth factors that may influence cancer cell growth the most.

And I don't even have a sweet tooth! In fact, I couldn't care less about desserts. But *had I known* just how dangerous it was, I never would have succumbed to a warm chocolate chip cookie or a hot fudge sundae on that cross-country flight, which was where I usually indulged, because I am not a dessert eater. Admittedly, it's hard to say no when they bake those cookies in the galley and the entire cabin fills with the sweet smell of freshly baked chocolate chip cookies. There's something comforting about it that teases the brain into saying yes when you ought to say no.

Willpower be damned!

Sometimes when the flight attendant came by with that silver tray, I closed my eyes tight, stuck my fingers in my ears, and sang "la-la-la-la-la" until she got the idea and tempted the next passenger with her dessert.

Had I known my life was actually in danger, it would have been so easy just to say no.

And now that I know, I cannot unlearn all of this.

But I can and will pass it along.

Going forward, I will share with others what I've learned about the signals that foods can send to your cells, how some foods put you at risk and others actually protect you.

I hope that as we go forward, doctors won't recommend the same treatment to every woman, just because that's the way it's always been done, if there is something new and more promising that better fits her particular needs. This is a field where research discoveries change detection, treatment, and survival rates week to week, month to month.

I think I'm a good example that trying a new regimen brought a good outcome.

I hope that as we go forward, we focus on treating the whole woman and not just the cancer.

I hope that doctors consider the long-term side effects that can result from the cancer treatment itself. It's easy for doctors and patients to focus on the more immediate and obvious side effects—like hair loss from chemo—and those unfortunately can and often do trump discussion about the long-term side effects that may occur after chemotherapy.

Finally, I hope that as we go forward, we can *focus on the doctor-patient relationship*. I am aware that I had the very best medical care available. I had it so good. However, in going public with my cancer journey and connecting with thousands of women who have shared their stories, I am also aware that not everyone has had the same quality care.

I was extremely fortunate to have incredibly wonderful relationships with my doctors. Each one helped me to understand my disease and all of the treatment options available to me.

I am making a promise here and now.

I will look for every opportunity to put myself in front of medical professionals and impress upon them the importance of their relationship with you and what you need from them:

1. To always share pertinent lifesaving information;
2. To probe and help you ascertain what your real risk factors for cancer may be;
3. To see that you get any and all ancillary tests that may be required, considering your health circumstances;

4. To explain all of the potential treatment paths available and why certain ones may better fit your needs;

5. To communicate with all other doctors treating you so that you get the best care; and

6. To remember that other, very important prescription: a caring and comforting smile and a word of encouragement that together we can beat this, and that you must stay positive and believe you can beat this, and that you may or may not have some lousy days, but in the end, you will survive!

My busy and hectic lifestyle required me to work with three different oncologists, and they all worked together *for my benefit.* That has shown me that once you are diagnosed with breast cancer, your chances for getting the best possible care are highest if all your health care professionals are involved in your diagnosis and treatment and work together. My doctors all helped me believe that I *could* win my battle, and I think this is so important to the success of your treatment. But you can never forget that you have to be your own advocate. You are in control of your health and, ultimately, your journey.

I didn't ask to get breast cancer. But I feel as though I have become a voice for women going through the battle. While I am not a medical doctor, I sure do feel like I've had a crash course in breast cancer! To be certain, I have the attention of many women going through this, and I am thankful that I can help to educate and inform them along their journey.

Had I known before my cancer diagnosis that there were so many roadblocks, frustrations, and discouragements in the battle against breast cancer I would have stepped up to the plate much sooner than I did. *Had I known* how great this need was, I would have—and feel I should have—tended to it sooner, to effect change and to become a voice for those who are in a predicament that requires someone to protect them, as I can now. Of course, it took getting breast cancer to fully understand the plight we all face: the bureaucratic red tape, the confusion, and the plethora of medical options left in our hands to decide on.

But when I did understand this, I knew I wanted to become a facilitator of information—to bridge the gap between the scary unknown and the necessary information every breast cancer patient wants and needs at every stage of her journey. The answer came to me in the form of an online television channel called *ALIVE with Joan*, which serves as a platform for cancer patients, survivors, their circles of care, and anyone looking to protect themselves from chronic disease. Information empowers and community inspires. I know from my own journey there is great value in both.

ALIVE with Joan was launched one year to the week after my breast cancer diagnosis and serves as a reminder every day that although it wasn't easy, I survived. And through *ALIVE with Joan* I want to encourage others to know they can survive too—and that others in the breast cancer community will be right there with them every step of the way.

I'd like to think that if my dad's up there, he's looking down at me, smiling, and saying, "You pick up the ball, baby doll, and run it into the end zone!"

I will, Dad.

I will.

Acknowledgments

I'd like to thank some very special people in my life who helped me through this past year and who made this book about my breast cancer journey possible.

I share my life journey with a most amazing life partner. I could not ask for a better copilot than my husband, Jeff Konigsberg. Thank you, Jeff, for your never-ending love and support and also for putting up with all of the hours that my eyes were glued to my computer screen as I wrote this book.

My three older daughters, Jamie, Lindsay, and Sarah have always been the lights of my life and they were there with me every step of the way throughout this journey. I am so proud of each of you. You have no idea how important it has been to have the three of you to turn to day after day. And to my younger children, Kate, Max, Kim, and Jack: thanks for helping to keep

Mommy strong and for telling me how cool I looked every time I took my wig off. Jeff and my children all surrounded me with their love and strength and I am so thankful to them.

Family members are often your pillars of support and your best cheerleaders and I thank Jeff's parents, Janey and Donnie, and his brother and sisters: Kip, Leslie, and Karen, and his extended family. I didn't come from a big family and you have all shown me how important unconditional familial love can be.

Friends are the family you choose. I'd like to thank close friends Elise Silvestri and Jill Seigerman who are like family to me—they've been remarkable listeners and have given me sound advice over the years. Every woman needs close girlfriends to vent to, to get a manicure with, to plan baby showers with, and in this case to pick out wigs with. I thank them for being there on the stormy days as well as for the exciting times. I thank them for their loyalty and friendship. And speaking of friends, I always know I can count on Michelle Dillingham—I'd say she's one of my oldest friends, but neither of us likes the word *old*. We will always be best friends, because we know way too much about each other.

I'd also like to thank Tara Girouard and Torie Sutterfield, two of the most loving people, whom I depend on each and every day to keep our family ticking and our children happy. Two sets of twins go four separate ways and it takes a village, and as a working and traveling mom, and this year a mom in cancer treatment, I couldn't be more thankful to have them living in my village.

I was very lucky to have such amazing medical doctors to guide me throughout this crisis: my oncologists Dr. Ruth Oratz, Dr. Tracey Weisberg, and Dr. Dickerman Hollister, and my breast cancer surgeon, Dr. Barbara Ward. I also want to thank my radiation oncologist Dr. Ashwastha Narayana and his amazing staff and also my radiologist Dr. Gail Calamari, who first found my tumor. Thank you too to Dr. Albert Knapp who has been my lead medical doctor for decades and who helped me put together my cancer team. And, deep gratitude to Dr. Cliff Hudis, who helped me find Dr. Weisberg in Maine when I needed to make the move.

My journey was made so much easier and kinder by some wonderful nurses: Beth Taubes in Dr. Oratz's office; Dawn Whitten and Jenny Jamaison in Maine; Joy Paul, Nancy Scofield, Mary Beth McFadden, Manny Acevedo, Lynn Carbino, in Greenwich; Heidi Malin, Carolyn Troy, Cindy Coyman, Dean Oliver in Radiology. You are my heroes. I thank you all for helping me to feel brave and strong, even when I wasn't feeling so brave and strong!

I would never have known to get an ultrasound and perhaps my cancer may have gone undetected for some time if I hadn't been given the lifesaving advice by breast cancer pioneer Dr. Susan Love several years ago during an interview we did together. I truly believe that interview saved my life, and for that I will be forever grateful to her.

A special thank you to my good friend, confidante, and personal trainer, Beth Bielat, who really is so much more than my fitness trainer. Throughout my treatment, Beth helped keep my head on straight, my feet moving, my breathing from the belly, and my emotions in control—and even brought me home-cooked meals many mornings in Tupperware to help me maintain my clean eating program. Beth has helped me change my health and wellness and has also been a wonderful partner in putting together Camp Reveille to help inspire other women to make the same commitment in their lives.

I never knew that nutrition could play such a huge role in my ability to make it through chemotherapy. Thank you Dr. Rob Zembroski for teaching me the ropes of clean eating and the value of how food can sometimes be your most important medicine.

Many thanks to Michael DeBerry, the head chef at Camp Takajo during the summer. Michael prided himself on whipping up dishes that complied with Dr. Z's "clean eating" requirements while still managing to taste great. We both had fun finding ways to add blueberries to couscous and veggies to give it a sweet zing. No sugar, No problem. With his help, I was able to maintain my clean eating throughout my cancer treatment and beyond.

Within days of making my diagnosis public I immediately heard from some of the top philanthropic leaders in breast cancer. Leonard Lauder, not only of Estée Lauder fame but also a loving husband to his late wife, Evelyn, who bravely battled breast cancer and founded the Breast Cancer Research Foundation with her oncologist Dr. Larry Norton. Evelyn lost a later battle with ovarian cancer. I thank Leonard for being a supportive presence throughout my journey. Nancy Brinker called me as soon as she heard the news of my diagnosis. Nancy's sister Susan Komen lost her battle with breast cancer at only thirty-six, and Nancy made a promise to her dying sister to do whatever she could to support others battling the disease and to find a cure. Her accomplishments have been beyond anyone's wildest imagination. I so admire her compassion and dedication to women everywhere and her determination.

The public telling of this story truly began with those first words spoken on the air at my alma mater, ABC TV. A very special thanks to Robin Roberts and my *Good Morning America* family for their kindness in helping me say those difficult words to America— that "I have cancer." Robin has been a friend to me when I needed it the most and I will forever be thankful for our sisterhood.

I'd also like to thank Kate Coyne, senior editor at *People* magazine for helping me make a huge life decision. It was her generous support and perseverance that helped me understand the importance of allowing myself to be photographed bald for their September 2014 cover in order to become a voice for thousands of others battling cancer.

A big thank you to Ruven Afanador, the amazingly talented photographer who took that photograph that would change my life forever. Thank you, Ruven, for helping me to bravely look into your lens with a positive attitude and the twinkle in my eye that I needed, to be able to pull off that cover shot. Also thank you to Freddie Leiba, who styled me that day and made me feel glamorous enough in my clothes that I could comfortably take off my wig.

And for that *People* shoot and far, far beyond, Emir Pehilj keeps me looking like a fresh, "ready for your close-up" Joan Lunden wherever I go, caring for both my makeup and my hair, or in the case of the last year—my wigs and peach fuzz. He makes it so much easier to put on a happy face in the midst of an unexpected crisis. Emir is such an amazing artist whom I greatly appreciate not only for his talent but for his loyalty as a friend.

When it comes to making the right decisions with respect to the press, you can't ask for anyone better than Stan Rosenfield. He not only helps me keep things in perspective but helps me approach everything with a sense of humor. Thank you, Stan, for stepping in when I need you most, but especially for being there when I faced making the difficult announcement of my cancer diagnosis. I can always depend on you to help steer me in the right direction.

I also want to thank my dear friend, confidant, and sage adviser, my attorney Marc Chamlin. He's been through a lot of chapters of my life over the past several decades, helping me navigate my path. I consider myself most fortunate to have Marc there with me on this journey.

I also want to thank Rick Hersh for his unwavering support throughout my journey.

And I absolutely must thank Richard Koenigsberg, my financial confidant, and his staff who helped me navigate through the financial minefield of my cancer care treatment.

To many, a cancer diagnosis can make them very nervous about their career, and I was no different. However, I have the good fortune to work with several wonderful companies that rallied around me throughout this challenging time. I am eternally grateful to Dr. Howard Murad and the entire Murad family for supporting me over the last year. And to everyone at A Place for Mom, especially Tracey Fitzgerald, thank you for your patience and your loving kindness.

Sometimes out of life's most challenging moments come our most precious experiences. That happened for me when my good friends at the *Today* show asked me to join them as a special correspondent for Breast Cancer Awareness. Joining them for #pinkpower was one of the most inspiring and rewarding weeks of television I've ever been a part of. I especially want to thank Matt Lauer, Hoda Kotb, and the entire *Today* show team who welcomed me so graciously and made me feel so comfortable, especially Debbie Kosofsky and her production staff. I will always be grateful for the warm reception everyone showed me at a time I needed it most.

By sharing my journey with the world I have had a unique opportunity to connect with thousands of

Americans who have been touched by cancer or illness in some way and who reached out to me via social media. I'd like to thank each and every one of you who motivate me to fight the good fight, as they say.

Elaine Capillo and Ali Barrella work day in and day out in my office on the many details that keep my life running, not only throughout the months of my breast cancer treatment but every day. Your dedication, hard work, and support are more appreciated than you will ever know.

I know I always love seeing behind-the-scene photos, so we kept a camera nearby all throughout this journey. Facing your mortality also prompts you to review your life and as I did so, I tried to find photos to reflect that as well.

I met Laura Morton, my coauthor, as I made my way through another life transition—taking charge of my health and my happiness as I turned forty. I had morphed from a sedentary, unhealthy, overworked mom of three to a fit, healthy, happy person and she was there to document it—first in *Workout America*, a fitness video, and then in the books *Healthy Cooking* and *Healthy Living*. Having lived through profoundly life-changing chapters with her, I knew that I would only want to share bringing this journey to life with Laura. Thank you, Laura, for caring as much about me and about my story as you do. I thank you from the bottom of my heart. I think it was meant to be that you would write about this breast cancer journey, for you shared a similar journey with your mother, who was taken from you at much too early an age by cancer. I am sure that she is looking down and is so very proud of you.

I want to thank my literary agent, Mel Berger, at William Morris Endeavor for his support and guidance. He shares my dedication to inform and empower women about breast cancer.

From our very first meeting, I knew that HarperCollins would be the perfect home for this book. My sincerest thanks to everyone on the HarperCollins team who has worked so hard to help create this book and for their efforts to get it into your hands. I especially want to thank my publisher, Jonathan Burnham, and I also want to thank Lisa Sharkey for sharing my passion for this project and for her guidance along the way. A special thanks to my editor, Amy Bendell; Alieza Schvimer; associate publisher Kathy Schneider; my PR and marketing team, Tina Andreadis, Katie O'Callaghan, Leah Wasielewski, Leslie Cohen; and art director Gregg Kulick, for his beautiful cover design.

And to my mom, Gladyce: I may not have faced this challenge with the same positive attitude had you not instilled in me at an early age the knowledge that I am strong and that nothing is impossible and that every cloud has a silver lining.

Finally, to my dad, my compassionate, dedicated, amazing dad, who I've always looked up to, and who I always wanted to make proud. I think of you each and every day and when I stand before all the different cancer organizations speaking, I am not standing there alone—you are right next to me, smiling proudly, which fills my heart with great joy.

About the Authors

Joan Lunden was the cohost of *Good Morning America* for nearly two decades, bringing insight to the day's top stories, from presidential elections to health & wellness. Her bestselling books include *Joan Lunden's Healthy Cooking*, *Joan Lunden's Healthy Living*, *Wake-Up Calls*, and *A Bend in the Road Is Not the End of the Road*. Joan speaks all over the country about health & wellness, inspiration, and success. Her online TV channel created for the breast cancer community, *ALIVE with Joan Lunden* (AliveWithJoan.com) and her website, JoanLunden.com, have quickly become go-to sources of information for women, bringing together experts on a myriad of relevant topics for today's woman. In October 2014, Joan joined NBC's *Today Show* as a

Special Correspondent for Breast Cancer Awareness Month.

Laura Morton is the author of more than forty books, including nineteen *New York Times* bestsellers. She has worked with Susan Lucci, Al Roker, Melissa Etheridge, Sandra Lee, and Justin Bieber among many others.